看起來都是那麼的理所當然，但太早說或做，別人就當我是瘋子。

　　這些年我花了一些時間在研究電商跟 AI，起始是因為我這二年創的小品牌 Pinwelife，需要一些材料，而我又不願意將組織工作變得複雜，所以開始找一些可以節省人力產出的解決方案。ChatGPT 對外公開發表開放試用時，我應該算是排很前面的使用者，雖然當時我對 AI 應用的架構已有所理解，但 ChatGPT 的表現，還是讓我感到驚奇，當然有時好笑。那時很多人都在挑戰它的能耐，最常看到的就是輸入自己的名字，然後要它寫介紹或解釋，當然就出現一些啼笑皆非的答案回應。我當時是這樣跟朋友說：「不是它不行，而是你不行」。不然試著輸入比爾蓋茲或華倫巴菲特看看，正確率超過 90%。很多誤會，歸因於理解錯誤。

　　「理解」AI，是我們今天要使用 AI 工具的第一步，要知其然，也要知其所以然。理解的太低，大材小用浪費資源；過度的期盼，就掉入不斷嘗試錯誤的迴圈。現在的 AI，只能夠稱為「弱人工智慧」，現在各家積極在研發的 AGI 通用型人工智慧（Artificial General Intelligence），才稱得上是強 AI（Strong AI），而這十年內必然發生。更簡單的定義強 AI，就是一個不會喊累、智商又超過 140 的「人」。而強弱之間主要的區別則在於「主動性」，弱智就是你問甚麼，它給甚麼（還不一定對），而強 AI 甚至能修正並強化你的需求，並可以主動地修正錯誤。但在強 AI 時代尚未來臨前，如何利用「弱人工智慧」來提高生產效率進而達到我們的目的，關鍵就在於明確定義需求。降低我們需求與 AI 理解的落差，進而生成一致的結果，「知己知彼」同樣適用於 AI 時代。

這本書是要寫給誰看的？在構思書本結構時，這是我問自己的第一個問題。

首先，我想推薦這本書給所有不會繪畫、不懂設計的人看。我自己算是美術設計科班出身，說天分嗎？也許有一點，但絕對不是那種大師巨匠等級的，至多就是還有點美感，而且我個人缺乏手繪能力，就是我們說的手殘黨，手繪能力基本上是無法學習的，更多是天賦，就像小時候或許曾經很羨慕看到同學在課本上隨手亂畫，大象是大象，兔子是兔子，而有些人永遠只會畫火柴人。現在，只要知道大象長甚麼樣子，任何人都可以畫出各種風格的大象。

　　再來我想跟所有創作者說，或許也該看看。我自己的國、高中同學，除了我比較離經叛道，最後從事的工作跟美術設計比較沒有關係，多數的同學，應該超過九成都是從事相關工作，職業藝術家，美術專業老師，室內設計師、插畫家、雕塑陶藝家、多媒體創作……各種藝術設計領域幾乎都有。當然我們那年代是從紙張、顏料、筆桿、調色盤，水袋、畫架、針筆、鴨嘴筆、雲形規、製圖工具、製圖桌一路走來的，而現在 AI 的應用，在同儕圈其實也是討論的風風火火，有人堅持情懷至上，手創無價；有人說 AI 創作不正經，當然也有人很開心地說「現在出設計圖快多了」。無論你是贊同、質疑或批評，在每個時代巨輪轉變之際，都必須面臨到價值觀的重新建立，然後才是工具、流程、組織的檢討。

　　最後這本書我想鼓勵所有的老闆或是高階經理人都要看，你們也許沒有空閒自己去嘗試這些東西，但請一定要知道現在工具可以做到甚麼地步，然後接下來該如何利用這些工具去調整、培訓現在的組織人力、重新定義工作流程等等。

人人都會 AI 繪圖

開啟斜槓人生金鑰匙
2000件生成作品＋完整提示詞

黃稟洲 Louis Huang ／著

suncolor
三采文化

作者序

AI 將打破階級的可能機會

老實話，一開始壓根沒想要出版這本書，我就是在嚐鮮、在玩一些讓自己開心的事，如果分享給朋友，他們看了也會開心那就很好。所以我想先交代一下自己的想法，也是這本書從無到有的過程。

幾個月前忽然接到三采文化張輝明董事長的電話，張董事長是我北士商廣設科的導師，交情互動超過 30 年。「老師好，什麼事？」、「錦忠（我以前的名字），我最近看到你在臉書分享的那些 AI 畫的圖，你應該趕快把這些整理一下，出書教大家。你要自己出版還是要三采幫你出都可以。」、「喔，好，我整理一下資料。就麻煩老師出版。」三句對話，就決定要寫這本書了。

我退伍之後就沒在從事設計相關工作了，一頭栽進資訊電子業，從行銷業務做起，一直到管理千人百億的企業，在職場發展上好像也還行，這 30 年的職場經歷，我算過得很精采，典型水瓶座怪咖的天性，總是做些旁人眼裡怪怪的事，但後來證明其實不太怪，我只是發現的比較早。我 1998-2000 年去矽谷開公司，那時候我就在做 B2B 電商黃頁，馬雲還沒開始。只是後來他撐住了，我們倒了。在那個數位相機剛出來只有 30 萬畫素的時候，我提了一個類似今天 youtube 的平台，希望大家分享上傳影音圖片資料，理想是如此美好，殘酷的現實是整個應用環境還不成熟，沒人理我們。「年輕人終究是年輕人」這老梗一直都適用。網路公司結束後，我自己去弄了一間出版社，出了第一本也是目前唯一一本台灣本土的網頁設計年鑑《網頁設計年鑑 2002》，在那個大家拼命做企業網站的時候，我已經在談 Usability/UI/UX。這些事現在

各種 AI 工具的出現，我真心覺得是送先天不足的人一場造化，更是打破階級的可能機會。與五感或四肢有高度相關的工作，成就高低幾乎是出生那天就決定了，無論是張大千、畢卡索對於事物的洞見與觀察入微，或是莫札特、郎朗這種超乎常人的聲音敏感度，都是天賦讓他們有機會成為一代巨匠，我並非說不需要努力，但努力的目的只是讓天賦得以展現。有些東西是練不來的，勤能補拙在需要天賦的項目上幫助有限。

AI 的問世，某種程度上彌平了一些（天賦）差距，如果說張大千、畢卡索這些大師巨匠是 100 分，那麼美術系教授可能有 60 水平，美工科第一名學生 40 分，那麼 AI 產出就已經具備了 40 分的水準，而這 40 分是可以普惠到任何人、任何人！即便是色盲或是無法拿筆的不便人士，都有機會可以得到這 40 分，而多數生活甚至商業應用，40 分已足夠；多些人，40 分已經滿足需求。

無論你是企業主、經理人或是社區媽媽，希望你們看完這本書後，我們同樣都具有 40 分的創作水平，而原本就已經是 60 分的專業工作者，可以不要再去做那 40 分的工作，將心力投入在創意，自我提升為 80 分的人才。這才是科技存在的意義。

技術與文化共舞：AI 繪圖時代的創作新思維

輔仁大學英文系副教授 劉雪珍

　　當代人工智慧 AI 技術蓬勃發展，AI 生成繪圖已成為設計界熱烈討論的議題，相關企業也紛紛強調員工應掌握 AI 繪圖的技能，Louis 的這本書出版得正是時候！這種人機協作的藝術新紀元在 21 世紀的今日即將啟航，幾乎是手把手地教普羅大眾如何運用 AI 創作。

　　我認識 Louis 超過 30 年。他是我輔仁大學英文系學生，可以說我一路看著他成長！畢業後他任職網路科技公司、開過公司，做過電商，不斷自學，拿了台大 EMBA 碩士，後來索性再攻讀博士。他機靈，愛好自由，喜歡探索與創作。他嗅到商機，知曉 AI 繪圖的強大在於它能以極快的速度生成令人驚嘆的視覺效果，並輔助設計師完成繁重的基礎工作。然而，它的局限性也同樣明顯，其本質仍是基於既有數據與算法的工具，無法真正理解人類情感與文化脈絡。文學與文化，則是人類歷史的沉澱，承載著無數代人的情感、哲思與價值觀。若僅仰賴 AI 繪圖而忽視文學文化內涵，藝術創作便可能淪為形式的堆砌，失去觸動人心的力量。

　　試想，一位設計師若沒有文學的涵養，如何能準確捕捉作品背後的故事與情感？而文化背景的理解，又如何影響作品中元素的選擇與呈現？例如，一幅 AI 生成的中國風圖案，若缺乏對中國詩詞與書法的了解，可能僅僅流於表面的山水與亭台，難以傳遞文化的深層韻味。Louis 在書中實證，若設計師具備深厚的文化修養，加上精準的提示詞，更能透過 AI 生成兼具傳統底蘊與現代感的作品。

　　融合 AI 技術與文學文化，能提升作品的質量，更能賦予設計更多層次與意義。這種融合思維如魚幫水、水幫魚：AI 提供技術支援，文學文化則提供靈感與深度。設計師可以利用 AI 快速生成初步的設計框架，然後以文學的眼光賦予它生命。同時，文學的情感表達也可透過 AI 的視覺化能力，打破文字的局限，觸及更廣泛的大眾。

　　在教育與職場環境中，我們提倡「技術與文化並重」的理念。對年輕一代來說，掌握 AI 技術固然重要，但培養文學與文化的素養同樣不可忽視。唯有在兩者之間找到平衡，未來的創作者才能不被技術所限，真正以人類的智慧與情感，創造出與眾不同的藝術與設計。

　　Louis 的這本書為我們帶來了無限可能，也提醒我們不要忘記文學與文化的價值。技術是工具，文化是靈魂，兩者的結合讓創作煥發光彩。我們不僅需要會使用 AI 的設計師，更需要具有文化底蘊、能以 AI 為助力的藝術家，為未來創作開創新的篇章。

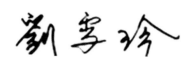

推薦序

重新定義你的能力，生成你想要的未來

台灣大學管理學院產學發展副院長 **謝明慧**

2024 的聖誕節剛過，你可能觀賞過世界知名可樂公司撥放的電視廣告。沒錯，即使是財力雄厚的大企業也利用新科技，完成了複雜的影片製作。一上片，罵聲多過掌聲：民眾普遍認為這是為了節省經費的投機取巧，但卻代表又有一批藝術家，甚至一整個行業被科技顛覆。

你我都身處這樣的時代：AI 技術日新月異，人們的恐懼也與日俱增；你可以選擇站在原地，默默等著被人工智慧大潮吞噬；也可以思考如何乘風而上，巧妙借力科技。但該如何踏出第一步呢？

Louis 自就讀台大 EMBA 起便展現對新興科技與商業應用的濃厚興趣，每逢新技術誕生，他總能迅速掌握核心並思考其本質。《人人都會 AI 繪圖——開啟斜槓人生金鑰匙》正是 Louis 探索與實踐的精華，副標題則是我和許多人的心聲。本書涵蓋了 AI 影像生成的應用情境及商業可能性，能讓初學者快速入門，也能讓企業主與經理人有機會在 AI 時代重定義起跑點，適合所有想在技術浪潮中重新定位個人或組織的讀者。

Louis 以其豐富的實作經驗與天生的藝術美感，將令人害怕的冰冷科技解釋得栩栩如生。讀罷方知：原來人工智慧並非工程師專利，AI 繪圖更是垂手可得的趣味創作！這是一個可以跨界轉職的斜槓時代，你需要領悟的是達文西的智慧，至於無以倫比的意象就請 AI 來幫忙生成吧！本書不僅是技術指南，更展現出 Louis 多元跨域的洞察力以及對生成式 AI 在商業應用上的深度思考。他以倡議者與實踐者的雙重身分，提出獨特見解，從工具導入、流程優化到組織革新，都能見到他縝密的思考脈絡，讓讀者理解將生成式 AI 落實於經營管理中的可行性。

身為商管領域的教育工作者，我對科技在教育與商業裡所扮演的角色始終感到好奇。誠摯推薦本書給所有對生成式 AI 感興趣或是徬徨的讀者，願它成為你善用科技的最佳起點。

目錄

PART 1　三分鐘進入 AI 領域

PART 2　關鍵提示詞及深度使用 Midjourney

PART 3　如何寫好提示詞

前言

AI 繪圖不僅是技術面向的提升，更是思考方式與創意流程的重塑

在這個數位經濟蓬勃發展的時代，人工智慧（AI）已不再只是前沿科技領域的艱深研究成果，更以多元而豐富的形式，深刻影響著我們的生活與工作模式。其中，AI 繪圖技術的興起，正迅速改寫著藝術創作與視覺設計的生態版圖。無論是商業品牌的形象重塑、藝術家的靈感啟迪、行銷專案的視覺策略，甚至是一般創作愛好者的日常美感探索，AI 繪圖正透過世代演進的演算法、強大的運算資源，以及豐富多樣的訓練資料庫，創造出前所未有的創意空間。

本書名為「人人都會 AI 繪圖—開啟斜槓人生的金鑰匙」，它的誕生基於一個清晰的信念：AI 繪圖不再是少數專家或科技愛好者的專利，而是所有人都能掌握的創新工具。透過本書，我們希望為讀者們提供一把關鍵「金鑰匙」，使您能夠以更寬廣的視角與更靈活的思維，跨越傳統職業分工的限制，拓展屬於自己的多元價值。對於想追求斜槓人生的人來說，善用 AI 繪圖工具，就如同掌握一項新奇的技能，讓您在不同領域靈活切換角色，創造更多元的職場與生活可能性。

在傳統繪圖與設計的語境中，創作的門檻往往不低：需要長時間的美術訓練、紮實的色彩與構圖理論、精湛的繪畫技法，以及對工具與軟體的熟練運用。然而，AI 繪圖技術的崛起，改變了這樣的生態。透過自然語言描述、關鍵字提示（Prompt）以及對風格、場景、主題的精準設定，我們只需透過言語與文字，便能讓 AI 產生無數風格多元、極具創造力的圖像作品。這不僅是技術的進化，更是創作模式與創意邏輯的根本轉型。AI 在此不僅僅扮演「工具」的角色，它同時是我們的「對話夥伴」，透過不斷的溝通與嘗試，AI 得以理解我們的構想，並不斷產生令人驚喜的作品。

本書不僅要帶領讀者了解 AI 繪圖的基本原理與操作流程，更希望透過多元案例與實務指南，引導讀者培養一種全新的美學敏銳度與創作思維。在這個過程中，我們將逐步探索 AI 繪圖的生成基礎，包括深度學習（Deep Learning）、生成對抗網路（GANs）、擴散模型（Diffusion Models）以及自回歸模型（Autoregressive Models）等技術的脈絡，協助您理解 AI 如何從龐大的影像與文字資料中萃取特徵，以近乎無中生有的方式建立出全新圖像。同時，我們也會詳細剖析當下最受歡迎的生成式 AI 平台與工具，例如 Midjourney、DALL·E（Copilot Designer, ChatGPT）與其他進階應用。在這些章節中，讀者將學到如何精準撰寫提示詞，控制影像的風格、主題、色彩、構圖與細節，並從 AI 的回饋中不斷修正指令，優化最終成果。

然而，學習 AI 繪圖不僅是技術面向的提升，更是思考方式與創意流程的重塑。過去，藝術創作的核心在於人類的直覺、美感經驗與獨特視角；現在，面對 AI 生成的作品，我們必須思考：在這種人機協作的情境中，創作者的價值究竟何在？本書將探討這個核心問題，並強調人類創意者不會被 AI 取代，反而能藉由 AI 的輔助，將原本的創作能量放大數倍。透過 AI 的快速產出，我們得以在短時間內生成大量的視覺素材，快速篩選並尋找靈感所在。人類的創意智慧、審美判斷與敘事能力，仍是畫龍點睛的關鍵，使得 AI 的生成作品能更貼近我們的創作意志與美學品味。值此，我特別邀請國內知名藝術家卓淑倩老師合作創作，也特別藉由這個過程與讀者分享人機協作的可能性。

❶ 卓淑倩老師原作：夢游星瀾 / 油彩 /91x72.5cm/2024

❷ AI 根據創作者之想法生成：卓老師所提供之創作理念作為提示詞

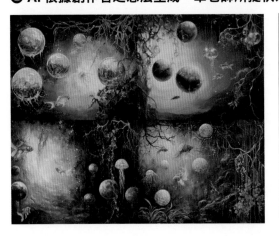

提示詞 ⇒ 一幅超現實的水下或星際場景的油畫，裡面充滿了綠藻狀的球形行星，其中一些是重疊的，發出黃綠色的光更接近梯度光源，而其他地方則是藍綠色的。類似花朵和藤蔓的懸掛植物出現在光源附近。六隻充滿活力的半透明金魚與背景融為一體，在十隻半透明水母的陪伴下優雅地遊動。幾株淺棕色直立的水生植物增加了垂直對比，而乾樹枝佔據了右上角，與無數橙色和粉紅色的小魚交織在一起。漸變光源從左上角開始，向右下方逐漸變暗。深背景是深藍色，帶有紫色和綠色色調，由光源的中間色調強調。這個空間是分層的，點綴著星星般的斑點，行星在其明亮的一面上有紋理、色彩繽紛的亮點，在其陰影的一面上有更薄、更光滑的油漆。氣氛空靈、夢幻、多維，具有令人著迷、超凡脫俗的氛圍。

這階段 AI 可根據提示詞掌握到表現的主題，但對風格的理解是與原作有明顯差距的。

❸ AI 參照風格生成：以原作作為風格參照

（上述提示詞最後再加上風格參照參數 --sref image url）

你可以比較這三者的差異，最重要仍是創作理念想法，也就是提示詞，表現技法可以快速地用 AI 分析原作產生，在不考慮最終成品的前提下，在螢幕顯示的數位影像幾乎可模擬出原作的五成以上。創作者可善加利用 AI 工具，或將此做為創作草稿，將概念先行視覺化。

對於廣大讀者而言，這本書像是一座橋樑，幫助您跨越技術的鴻溝，進入 AI 輔助創作的嶄新領域。無論您是剛剛入門的初學者，還是已有基礎的專業設計師與藝術家，本書都將提供適合您的內容與學習路徑。我們將從最基礎的原理講起，讓讀者認識 AI 繪圖的核心關鍵：為何 AI 能從文字描述中「想像」出圖像？AI 如何理解風格與美學？在不同生成模型之間，有哪些特性與局限？透過對這些問題的深入探討，您將能從根本上理解 AI 繪圖的運作原理，不僅止於應用工具，更能融會貫通，靈活運用。

在技巧應用的層面，我們將細緻介紹提示詞撰寫的策略與訣竅。提示詞不僅是語言描述，更是一組「創作指令」，透過精確的關鍵詞、適切的風格描述與細膩的表達方式，您將能引導

AI 在特定的美學框架中創造出獨特且鮮明的作品。範例包含了從寫實人像、抽象藝術、品牌識別、概念插畫到產品原型視覺化的各種應用案例，我們不僅會告訴您如何產出某種風格的畫面，也會帶您拆解、分析與反思，讓您在一次次的試驗中掌握提示詞的奧妙。

本書的另一個特點是強調人機協作的動態流程。在創作過程中，AI 的回饋常常是不可預期的，有時候產出過於驚世駭俗，與預期落差甚遠；有時則能帶來意料之外的美感與靈感。面對這樣的「不確定性」，本書將幫助您理解如何善用試錯的歷程，不斷調整提示與條件，引導 AI 進入更精準的創作範疇。如此一來，創作不僅是單向的技術操作，更是一個雙向溝通、隨機應變的有機過程。這也使得創作本身變得豐富而有趣：我們不再是孤獨的藝術家，而是與 AI 並肩作戰的夥伴，在無限的創意空間裡彼此激盪、成長。

除了技術與創意思維，本書也嘗試觸及 AI 繪圖所帶來的社會與產業影響。隨著 AI 技術的門檻不斷降低，越來越多人將有機會以 AI 繪圖作為自我實現或職業發展的工具。在行銷、品牌設計、影視娛樂、遊戲美術、教育訓練、室內設計與服裝創意等領域，AI 繪圖正迅速成為新一代的視覺核心。「斜槓」的概念在此時展現出強大的生命力：一個擅長內容行銷的人，或許能透過 AI 繪圖提升品牌的視覺溝通品質；一位在藝術創作上有一定基礎的設計師，也能藉由 AI 生成工具，快速嘗試不同的視覺風格，為客戶提供更豐富的選擇。這樣的跨界與斜槓，不僅拓寬了個人的職業發展道路，更讓創作產業本身變得更加多元而蓬勃。

我們也必須面對 AI 生成影像所帶來的挑戰與責任問題。例如，AI 在處理特定題材時可能產生偏差，或在合成圖像時出現細節失真，甚至在真實與虛擬的邊界引發道德與法律爭議。本書將提醒您留意這些議題，並提出對應的策略與思考方式。透過了解 AI 的極限與盲點，能更審慎地使用這項技術，確保創作方向合乎倫理、尊重版權，並珍惜人類勞動智慧的獨特價值。

「人人都會 AI 繪圖—開啟斜槓人生金鑰匙」並不是一本技術教材的冷硬說明書，它更像是一盞指引您前行的明燈。透過對 AI 繪圖的深入探索，您將看見創作世界正在產生翻天覆地的變化：從一開始單純的「工具導向」，逐漸邁向「人機對話」、「創意增幅」的綜合過程。在未來，學習 AI 繪圖就如同學習一門新的語言，一個理解未來創作生態、通達跨界整合的必備能力。誠摯邀請您與我們一同展開這趟旅程。翻開本書，您將進入一個人機協作的藝術新紀元。在此，AI 不是冷漠的程式碼集合，而是激盪靈感的創作夥伴；在此，創作者的視野不再受限於個人技能的瓶頸，而是能依靠 AI 快速試驗多種可能，將創意能量放大到極限。讓我們共同擁抱 AI 時代下的創意浪潮，以 AI 繪圖為關鍵，開啟您的斜槓人生新篇章，將興趣、專長與事業巧妙融會，創造出屬於您獨一無二的多元未來。

機甲生肖 - 鼠

PART 1

三分鐘
進入 AI 領域

認識生成式 AI 影像

▶ 關於技術原理

AI 影像生成技術主要依賴於深度學習中的生成對抗網絡（Generative Adversarial Networks, GANs）和自回歸模型（Autoregressive Models）。這些技術使得 AI 可以從噪聲＊中生成逼真的影像，並且可以根據給定的文字描述來創建具體的圖像。以下是主要技術及其理論架構的詳細介紹：

■　噪聲：通常指的是數據中的隨機或無意義的變異，它可能會影響 AI 模型的性能。

一、生成對抗網絡（GAN）

GAN 由 Ian Goodfellow 等人在 2014 年提出，主要由兩個神經網絡組成：生成器（Generator）和判別器（Discriminator）。這兩個網絡通過對抗學習來提升圖像生成的質量。

● **生成器（Generator）**：生成器從隨機噪聲（通常是多維高斯分佈）中生成假圖像。生成器的目標是創造出足夠逼真的圖像，讓判別器無法區分真假。

● **判別器（Discriminator）**：判別器的任務是區分真實圖像（來自訓練數據）和假圖像（來自生成器）。判別器的目標是最大化判斷真實圖像為真，假圖像為假的準確性。

兩者通過對抗訓練（Adversarial Training）達到平衡，最終生成器能夠創造出極具真實感的圖像。

舉一個例子，讓你快速理解「生成對抗網路」：

想像一個場景：

你是一個正在學習繪畫的藝術家，而你有一個非常嚴格的老師。

- **你（生成器）**：你的目標是創作出足以以假亂真的畫作，讓老師誤以為是出自大師之手。
- **老師（判別器）**：老師的任務是仔細審視你的作品，分辨出哪些是你畫的，哪些是真正的大師作品。

訓練過程：

一開始，你可能畫得很糟糕，老師一眼就能看出破綻。但隨著你不斷練習，觀察大師作品，你的技巧越來越好，畫作也越來越逼真。同時，老師也不斷提高判別能力，能發現更細微的差異。

- 你不斷嘗試畫出更好的作品來欺騙老師。
- 老師也不斷提高鑑別能力，找出你畫作中的瑕疵。

最終結果：

經過長時間的訓練，你可能達到一個境界，你畫的作品已經非常逼真，甚至連老師都難以分辨真假。這時，我們可以說你（生成器）已經成功地學會了如何創造出以假亂真的畫作。

● **GAN 的精髓：**

- 生成器和鑑別器相互對抗，共同進步。
- 生成器努力創造更逼真的數據。
- 鑑別器努力提高分辨真假數據的能力。

● **GAN 的應用：**

生成逼真數據的能力，讓 GAN 在許多領域都有廣泛應用，例如：

- **圖像生成**：生成逼真的人臉、風景、藝術作品等。
- **圖像轉換**：將照片轉換成不同風格，如將白天照片轉換成夜晚照片。
- **數據增強**：為機器學習模型生成更多訓練數據。

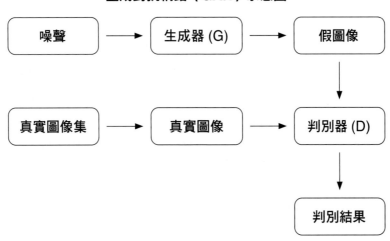

生成對抗網路（GAN）示意圖

二、自回歸模型（Autoregressive Models）

自回歸模型是一種生成模型，用於逐步生成數據，每一步都依賴於前一步的輸出。這些模型可以應用於圖像、文本、語音等多種領域。在圖像生成中，常見的自回歸模型包括 PixelRNN 和 PixelCNN，它們都是逐個像素地生成圖像。

● **PixelRNN（RNN）**：利用循環神經網絡（RNN）來逐像素生成圖像，每個像素的生成依賴於之前所有生成的像素。這種方法計算成本較高，但適合生成高質量的圖像。

● **PixelCNN（CNN）**：利用卷積神經網絡（CNN）來逐像素生成圖像。雖然也是逐個生成像素，但其並行計算的特性大大提高了效率。PixelCNN 使用掩碼卷積（Masked Convolution）來確保每個像素的生成只依賴於之前已生成的像素。這樣可以控制生成順序，從而保持圖像生成的連續性，同時相比 PixelRNN 提高了計算效率。

自回歸模型（PixelRNN/CNN）示意圖

三、變分自編碼器（Variational Autoencoders, VAEs）

變分自編碼器通過學習數據的概率分佈來生成新數據。VAE 由編碼器（Encoder）和解碼器（Decoder）組成，並且引入了一個隱變量（Latent Variable），使得生成的圖像更加多樣化。

● **編碼器（Encoder）**：將輸入圖像編碼為隱變量的概率分佈。
● **解碼器（Decoder）**：從隱變量的概率分佈中生成新的圖像。

這些技術結合起來，使得 AI 能夠從文本生成高質量的圖像，並且能夠根據具體需求進行自適應的生成。這些理論架構和技術應用為現代 AI 影像生成打下了堅實的基礎。

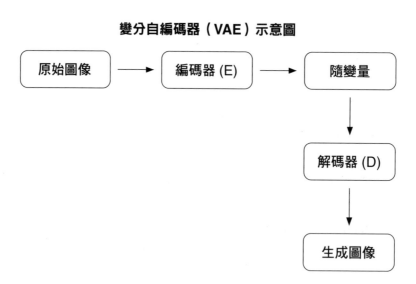

變分自編碼器（VAE）示意圖

▶ AI 生成的黑箱作業與幻覺

許多人對於 AI 的工作原理仍感到困惑，尤其是其「黑箱」（Black Box）運算性質和可能出現的「幻覺」（Hallucination）現象。接下來，以淺顯的例子說明：

一、AI 生成的黑箱作業

● 什麼是黑箱（Black Box）？

在科學和工程領域，「黑箱」指的是一個系統或裝置，我們只知道其輸入和輸出，但對其內部運作機制一無所知。在 AI 中，特別是深度學習（Deep Learning）模型，複雜的演算法和大量的參數使得內部過程難以理解。

● 為什麼 AI 是一個黑箱？

AI 模型，尤其是神經網路（Neural Network），通過大量資料的訓練（Training），學會在輸入和輸出之間建立複雜的關聯。然而，這些關聯並非通過人類可讀的規則，而是通過調整數百萬甚至數十億的參數。因此，我們很難解釋模型是如何得出某個特定結論的。

● 淺顯例子

想像一個魔法盒子，你往裡面投入一個蘋果（Apple），盒子會吐出一個橘子（Orange）。你知道輸入和輸出，但不知道盒子內部發生了什麼。這就類似於 AI 的黑箱作業。

二、AI 的幻覺（Hallucination）

● 什麼是幻覺？

在 AI 中，幻覺指的是模型生成了與事實不符或不存在的資訊。大型語言模型（Large Language Model, LLM）如 ChatGPT，有時會「自信地」提供錯誤的答案。

● 為什麼會產生幻覺？

AI 模型通過學習大量的文本資料，來預測下一詞或生成回答。但它並不具備真正的理解能力。當缺乏足夠的上下文（Context）或面對未見過的問題時，模型可能會「編造」答案，以符合語言模式。也因此，我們即使在 AI 影像生成工具中，輸入任何無意義的字，系統仍會提供生成影像。但並不表示這些無意義的提示可以驅動或控制 AI 生成，這是目前許多使用者的誤解。

在影像生成過程中，我們會常發覺一些不合理之處，例如：肢體動作、器官或其他物件的細節、物體相對比例等等，這些都是常見的「不合邏輯」，也就是目前 AI 產生的幻覺，其根本原因還是在於模型建構的完整性，又如目前多數的 AI 生成工具都無法順利生成中文字，這不是技術的問題，而是目前尚未有單位投入華語資料的模型建立。

● 淺顯例子

就像一個小孩在不知道答案的情況下被問到問題，他可能會憑空捏造一個回答。例如：問他「天空為什麼是藍色的？」，他可能會說「因為天空喜歡藍色。」這就是 AI 可能產生幻覺的類似情況。

三、應對黑箱和幻覺

● **提高模型的可解釋性（Explainability）**：研究者正在開發可解釋的 AI（Explainable AI, XAI），使模型的決策過程更加透明，幫助我們理解其內部機制。

● **加強資料驗證（Data Validation）**：在使用 AI 生成的內容時，需進行事實核查（Fact-checking），以避免被幻覺誤導。

● **人機協作（Human-AI Collaboration）**：將 AI 作為輔助工具，由人類來最終判斷和決策，可以充分發揮 AI 的優勢，避免其缺陷。

理解 AI 的黑箱作業和幻覺現象，有助於我們更理性地看待和使用人工智慧技術。通過淺顯的例子，我們認識到 AI 的強大之處，也看到了其局限性。只有在瞭解其工作原理的基礎上，我們才能更好地應用 AI。

上述內容或許對於非資訊理工的人較難以理解，簡單來說，**生成式 AI 影像，就是可以讓使用者通過簡單的文字描述（自然語言模型），就能創造出豐富多彩的圖像**。這項技術不僅提高了創作效率，還為各行各業帶來了新的創作可能性。

透過不斷學習和實踐，你將能更快掌握這項技術，創造更加驚豔作品。本書主要的目的，當然不是也無法教會大家去製作 AI 工具，而是希望大家能善用 AI 工具，增添生活樂趣，提升工作效率。

AI 影像生成可用在哪？

▶ 攝影的應用

AI 生成影像技術對商業攝影產生了很大影響。這項技術讓任何人只需簡單描述就能創作圖像，這在過去是難以想像的。它不僅提高了創作效率，還促進了視覺媒體的多樣性和創新。然而，這也引發了關於版權和原創性的討論。隨著 AI 生成影像變得更真實且普及，可能會使專業攝影師的需求減少，並可能導致圖像變得過於相似，缺乏獨特性。

儘管如此，許多藝術家和攝影師開始探索如何利用 AI 作為創意工具。例如，美國著名藝術家 Laurie Simmons 使用 AI 生成器創作了新作品，這些作品繼續探討她之前作品中的家庭和心理主題，並擴展了她的女性主義觀點。2023 年德國藝術家 Boris Eldagsen 特意以 Open AI 的 DALL.E 2 創作的黑白圖片《PSEUDOMNESIA | The Electrician》參加 2023 年 Sony 世界攝影大賽（Sony World Photography Awards）並獲得公開組創意類別首獎，雖最後拒絕領獎，但也凸顯了 AI 生成與傳統創作模糊的界線。整體來說，AI 被視為幫助藝術家和攝影師實現創意的工具，而不是取代他們的創意。

Laurie Simmons 使用 DALL-E. 所生成的影像
Source from：https://news.artnet.com/art-world/artists-ai-creativity-2340982

獲獎的《THE ELECTRICIAN | The Electrician》其實是 AI 生成。

AI 扮演的角色

◆ **虛擬攝影師**

AI 可以模擬攝影師的視角和技術，拍攝出各種風格和主題的照片。無論是風景、人像還是建築攝影，AI 都能根據提示詞生成高質量的圖像，滿足不同的攝影需求。

生成範例｜城市落日

想法 ⇒ 讓 AI 生成一幅日落時分的城市景觀，照片中包含高樓大廈、河流和倒影。

提示詞 ⇒ A cityscape photo at sunset, featuring tall buildings, a river, and reflections. --ar 16:9 --style raw
生成一幅日落時分的城市景觀照片，包含高樓大廈、河流和倒影 --ar 16:9 --style raw

◆ 後期製作助手（AI 增生）

在影像後期製作中，AI 能幫你進行照片修圖、色彩校正和加入特效。這不僅提升圖像的質量，還大大縮短後期製作的時間，讓攝影師能專注在創作部分。

AI 影像後製與傳統 Photoshop 後製在技術、效率和創作自由度上存在顯著差異。生成式 AI 影像後製能夠自動識別和處理圖像中的元素，如人臉、背景、物體等，並根據預設的算法和學習到的模式進行自動修飾和調整。相比之下，傳統的 Photoshop 後製依賴於使用者的專業技能和手動操作，如剪切、拼貼、調色、修飾等，依賴於使用者的創意和技術來實現最終效果。

總結來說，生成式 AI 影像後製適合快速生成和批量處理，而傳統 Photoshop 後製則在精細控制和創意自由度上更具優勢，使用者可以根據具體需求選擇最合適的工具和方法來完成影像後製工作。

◆ 創意引導

AI 技術能夠大幅提升攝影師的創意過程，不僅可以幫助他們激發新創意靈感，還能生成多樣化的圖像風格，讓攝影師在不同的藝術表現形式中進行探索。透過 AI 生成的圖像，攝影師可以嘗試新的拍攝角度、光影處理以及構圖方式，進一步豐富其創作視野和技巧。

此外，這種技術在概念圖創作、攝影作品集製作及視覺故事的敘述上，特別有用。有助於攝影師更快速地實現他們的創作目標，並展示出更具視覺衝擊力的作品。

以下我們使用 Adobe Firefly 進行簡單的示範：

❶ 準備要後製的影像。

❷ 選取「新增」功能，並將欲新增的區域快速選取出來，在提示詞對話框輸入「一隻巨大的筆」，按下產生。

❸ Firefly 會根據選取區域及提示詞，新增提示詞所描述之物件在選取區域。

使用 Midjourney 網頁版進行增生或修改：

❶ Edit 使用者介面，使用者可選擇輸入圖片連結或是上傳圖檔進行編修。

❷ 上傳了一張蒙娜麗莎的圖檔，選取左側橡皮擦工具，調整筆觸大小，將欲修改的範圍圈出。

❸ 在上方提示詞欄位輸入「抱著一個新生兒」，等待數秒後影像生成。

同理這樣的作法可以大幅應用在商務上，例如產品展示，隨意置換背景空間等等。而 AI 增生最大的好處就是會根據原始的影像去調整光線、材質、氛圍等等，相較於傳統的影像後製軟體，這是非常大的進步。

旅遊景觀

生成範例｜在湖邊跳舞的女人

想法 ⇒ 生成一幅旅遊景觀照片，包含景點與人物互動。既有現實感又充滿藝術性。

提示詞 ⇒ A blonde beauty dressed in a lace evening gown dances by the lake near the Angkor Wat temple. Her gown flutters in the gentle breeze, while the lights reflect on the water, adding a touch of modern luxury to this ancient setting. Her graceful and airy dance contrasts vividly with the solemnity of the surrounding ancient architecture, resembling a splash of vibrant color in the tunnel of time, evoking a sense of longing and admiration. Spectators may watch her quietly or be drawn in by her dance, filling the scene with a mysterious and enchanting atmosphere. in the style of Geoffrey Beene --style raw --s 1000 --v 6.0 --ar 3:2

一張真實的照片，身穿蕾絲晚禮服的金髮美女在吳哥窟寺廟附近的湖邊跳舞。 她的禮服，微風徐徐，燈光倒映在水面上，為這古老、輕快的舞蹈增添了現代奢華 --style raw --s 1000 --v 6.0 --ar 3:2

試想在過去如果要拍攝一張這樣的照片，所花費的人力、物力、時間等，是多麼的巨大，而今天只要善用工具平台，成本幾乎可以忽略，便可得到初步的效果。

操作解說：

❶ 輸入上面提示詞後，MJ 快速地生成了四張影像。

❷ 這階段每個人的選擇會不盡相同，比較符合我預想的畫面是第二張，湖中有燈光倒影，所以我們再針對第二張進行修改。選擇〔 V2 〕按鍵，會生成類似的四張影像。

❸ 選擇〔 V4 〕按鍵，會生成類似的四張圖像。

❹ 最後選擇第二張作為生成結果。選擇〔 U2 〕按鍵，就會得到單張影像。

生成範例 | 在湖邊跳舞的女人之延伸

想法 ⇒ 對前面吳哥窟的風景生成不滿意,想要將場景切換到埃及金字塔,甚至再賦予一些超現實的元素。

提示詞 ⇒ ealistic photo of Capturing the romantic rendezvous of Ramesses II, the powerful pharaoh of ancient Egypt, and his beloved queen Nefertari. Set against the backdrop of the majestic temples of Abu Simbel, depict the couple strolling hand, the Nile River --style raw --ar 4:3 --v 6.0

古埃及強大法老拉美西斯二世和他心愛的王后奈菲爾塔麗的浪漫約會的真實照片。以阿布辛貝雄偉的寺廟為背景,描繪了這對情侶牽手漫步,尼羅河。 --ar 4:3 --style raw

商業廣告拍攝

有些公司利用 AI 生成的圖像來創造引人注目的營銷活動,這些圖像能夠根據特定品牌的風格和需求進行定制,從而提供獨特的消費者體驗。

生成範例｜高級手錶的產品照

想法 ⇒ 生成一張高端手錶的廣告照片,展示其精細的設計和奢華的質感。

提示詞 ⇒ product, in tricate internal mechanical structure of a precision watch, tourbillon movement and ruby jewels, Hasselblad
產品照片,精密手錶複雜的內部機械結構,陀飛輪機芯和紅寶石珠寶,哈蘇相機。

創意攝影

在一些創意產業,如遊戲和電影製作中,AI 生成的圖像可被用來創建更加逼真的背景和角色,這不僅提高了生產效率,還增強了觀眾的沉浸感。

AI 生成圖像的商業應用範圍極其廣泛,涵蓋了廣告創意、產品設計、品牌形象塑造、數位內容創作等多個領域,極大地改變了企業和創意專業人士的工作方式。然而,隨著 AI 生成技術的進步,與之相關的版權和原創性問題也隨之浮現。由於 AI 生成影像的質量越來越逼真,這可能導致專業攝影師的需求下降,並引發市場上影像同質化的現象,進而削弱了影像的獨特性和多樣性,對整個創意產業的影響不容小覷。

生成範例｜魔幻感的森林城堡

想法 ⇒ 生成一幅奇幻世界的場景照片,包含神秘的城堡和魔法森林。

提示詞 ⇒ a dark forest with a witch casting spells, Witchcore
一片黑暗的森林,有女巫施展咒語,Witchcore

■ Witchcore 等核心風格的用法,將在 PART3 有詳細的介紹。

虛假影像製造

此外，AI 生成影像不僅在商業領域中具有重要影響，還對社會和文化產生了深遠的影響。AI 生成的圖像和影片可能被不當使用於製作虛假新聞、誤導性內容，這對於公共信息的真實性、社會信任度和整體資訊生態構成了潛在威脅。因此，在應用 AI 生成影像技術時，社會需要承擔起監管和道德責任，確保這些影像的真實性和合法性，避免技術濫用所帶來的負面影響。

目前，AI 技術已經達到可以生成幾乎無法辨別真偽的圖片和影片的水平，這對創作者和訊息接收者提出了更高的智慧與警覺性。在使用相關工具時，不僅要充分理解其潛力，也要意識到其可能帶來的風險，並做出負責任的選擇。

總體而言，AI 生成影像技術對於商業攝影及其他創意產業的影響是一把雙刃劍。它在帶來創新和效率提升的同時，也對傳統攝影、版權保護以及信息可信度提出了新的挑戰。未來，這一領域的發展將取決於技術、法律和市場之間的相互作用，以及我們如何在保護創作者權益與鼓勵創新之間找到平衡。隨著 AI 技術的持續進步，將有更多關於如何在保護創作者權益和鼓勵創新之間找到平衡的討論，以及如何妥善管理這些新工具所帶來的機遇與風險的討論將愈來愈重要。

生成範例│會生蛋的蝙蝠

想法 ⇒ 創意可以天馬行空，不可能的事可以請 AI 做。讓哺乳類動物蝙蝠也生蛋。

提示詞 ⇒ a real photo of Two big bats, with many small eggs around them. some of eggshells are cracked, and the heads of bat babies popped out. The background is a tunnel. It's with high resolution ab definition in very detailed. --ar 4:3 --style raw --v 6.0

兩隻大蝙蝠的真實照片，周圍有許多小蛋。 有的蛋殼破裂，蝙蝠寶寶的頭探了出來。背景是坑洞。 --ar 4:3

▶ 繪畫藝術設計的影響

AI 影像生成技術不僅對攝影領域帶來深刻影響，對於傳統藝術與藝術創作的衝擊更為顯著。AI 技術的出現，如同一把雙刃劍，既可以為藝術創作帶來無限可能，也可能威脅到傳統藝術的地位。正如「水能載舟，亦能覆舟」這句古諺所說，時代的巨輪已經不可逆轉地向前推動，我們既不應盲目吹捧這項技術，也不應過度抗拒。以更務實的態度來迎接這一劃時代工具的誕生。

創作流程的改變

AI 技術正徹底改變藝術家的創作流程，從一開始的靈感獲取到實際的創作過程，影響無處不在。

● 靈感和概念生成

AI 能夠快速生成大量圖像和設計概念，這對於藝術家在初期構思階段非常有幫助。像 Midjourney 和 DALL·E 等 AI 工具，可以根據簡單的文字描述生成詳細的視覺作品，特別適用於概念藝術和設計領域。這些工具能夠根據輸入的關鍵詞或簡單的草圖，生成多種風格和主題的影像，幫助藝術家能在短時間內獲得豐富的靈感，開拓創作的可能性。

● 技術輔助

AI 技術還可以幫助藝術家完成一些繁瑣和技術性的工作，從而讓他們能將更多時間和精力投入到創意表達上。例如，Adobe Photoshop 中的 AI 工具可以自動修復圖像、填補空白、進行顏色校正，甚至可以模仿特定藝術家的風格進行繪畫。這些技術大幅提升了創作效率，減少了手動修圖的時間，為藝術家提供了更大的創作自由。

藝術家身分和角色的轉變

隨著 AI 技術發展，藝術家身分和角色也跟著轉變。在以前，藝術家被認為是創意和技藝的化身，但 AI 技術日漸普及，使得更多人可以創作出高質量的影像，對藝術家獨特性的認知提出了挑戰。

● 創意民主化

AI 工具的普及使得任何人都可以創作出視覺藝術作品，這在一定程度上削弱了傳統藝術家的獨特地位。如今，非專業人士只需要使用 AI 工具，就能創作出媲美專業藝術家的作品，而這種「創意的民主化」在許多人看來，是值得推廣的觀念。

過去，藝術創作往往被「天賦」以及「特殊（專業）教育」所壟斷，就像我國中念的美術教育實驗班，高職念的廣告設計科。然而在今天，若暫且不論藝術價值，AI 生成的作品質量已經超過了許多受過專業訓練的美術科班生，或許「全民藝術家」的時代正在到來。

● 協同創作

許多藝術家開始與 AI 合作，將其視為創作夥伴而非競爭對手。例如，一些藝術家使用 AI 生成的初稿作為基礎，再進行進一步的藝術加工，這樣既保留了人類藝術家的創意，又利用了 AI 的高效特點。例如，著名的人工智慧藝術家 Sougwen Chung 與 AI 協作，跳脫出 AI 僅為工具的概念，視其為創作的合作夥伴。她的作品充分展示了藝術、技術

Sougwen Chung 與 AI 機械手臂共同創作。source from https://sougwen.com

和科學之間的動態交匯，探索了人類與機器協同創作的無限可能。

AI 在藝術領域的角色已從單純的工具，轉向富有主動性的「合作者」。許多藝術家在概念生成、資料處理與技術執行的流程中，引入 AI 作為參與者。例如，音樂家 Holly Herndon 在專輯《Proto》中採用 AI 合成聲音系統「Spawn」，以探索人聲與 AI 聲響之間的交融，最終呈現出前所未有的音樂形態。人機共創常具備「非線性、不可預測」的藝術特質。AI 與藝術家在創作過程中互動學習，形成了高度動態且開放的合作模式（Manovich, 2021）。這不僅推動藝術的邊界往更廣闊的方向拓展，也改寫了藝術家與作品之間的傳統角色分工。

藝術市場的變化

AI 技術對藝術市場也帶來了顯著影響，包括藝術品的生產、銷售和收藏方式。

AI Generated,《Edmond de Belamy》by Obvious, 2018

● 數位藝術市場

AI 生成的數字藝術品正在成為藝術市場的重要組成部分。NFT（非同質化貨幣）技術為數位藝術品提供了買賣和收藏的新方式，澈底改變了傳統藝術品的交易模式。例如，Beeple 的數字藝術作品在拍賣行以數百萬美元的高價成交，顯示了數字藝術市場的巨大潛力和投資價值。這一趨勢標誌著數位藝術品不再僅僅是數據文件，而是具有經濟價值和收藏意義的資產。2018 年，法國藝術團隊 Obvious 利用 GAN 技術創作的《Edmond de Belamy》以古典油畫風格呈現，最終在佳士得（Christie's）拍賣行以約 43.2 萬美元的價格成交，成為 AI 藝術發展的里程碑。

● 藝術品的價值判斷

AI 生成藝術品的價值判斷標準仍在探索中。傳統藝術品的價值往往基於藝術家的名氣、作品的獨特性以及其歷史背景。然而，AI 生成的藝術品如何定價和評估仍是市場面臨的新挑戰。藝術市需要建立新的評估標準，以衡量 AI 藝術品的價值，這可能涉及創意性、技術性、市場需求及與人類藝術家的合作程度等因素。隨著 AI 生成藝術品的增多，如何在藝術市場中確保其價值的公平評估，將成為未來的重要議題。

藝術教育的革新

AI 技術在藝術教育中的應用正逐步改變傳統的教學模式，提升學生的技術掌握和創意表達能力。

● 輔助教學工具

AI 工具在藝術教育中作為輔助教學工具，能夠幫助學生模仿並學習不同藝術家的風格，並且能夠提供即時的反饋，這對於初學者尤為有用。例如，DeepArt 等 AI 應用可以根據名著名畫作的風格對學生的作品進行即時轉換，這不僅幫助學生更好地理解不同風格的特徵，也能激發他們的創作靈感，讓學習過程更具互動性和趣味性。

● 美術教育的重新思考

隨著各種工具及科技的日新月異，在每個階段都必須要重新定義美術教育的方式，例如早期的攝影教學，從傳統的手動對焦到自動對焦，對於攝影師的技巧要求就不同；從底片相機到數位相機，過去重要的暗房工程理論及實務，就被數位影像後製處理所取代；各種方便的字體與電腦輸出，手繪 POP 海報也就逐漸淡出應用市場。

同樣的，AI 影像生成的興起，在美術教育上需要進行甚麼樣的調整及觀念上的重構，也會是所有美術教育作者必須正視的課題。前一陣子台灣著名的美術教學學校復興商工美工科成果展，首獎得獎作品便借助 AI 生成，我們先忽略比賽規則，純以結果論，便是 AI 的產出勝過手繪，更超越了評審老師的眼光。值此，我們更需要去思考，AI 影像生成在美術教育中所扮演的角色，可以是過程中輔助的工具，也未嘗不能是最終的結果。

手繪作品具有至高無上的情懷及藝術價值是無庸置疑的，但若不考慮**心理性價值**，純以實用或商業效率來看，AI 生成是否可取代過去的手繪工作？如同現今已沒有人在畫電影看板一般。

● 虛擬工作室

虛擬現實（VR）和增強現實（AR）技術使得學生可以在虛擬環境中進行創作，這為藝術教育提供了更多可能性。學生能在虛擬畫廊中展示和分享作品，與其他學生和老師進行即時互動，這不僅提升了學習效果，也激發了學生的創作熱情。虛擬工作室的引入，讓藝術教育不再受限於物理空間，提供了一個更為靈活和開放的學習環境。

AI 來勢洶洶，該如何面對？

面對 AI 生成影像技術的發展，傳統創作者需要調整心態，正確使用 AI 工具，以充分發揮其潛力。

擁抱技術

學習新技術：創作者應該主動學習和掌握 AI 影像生成技術，了解其基本原理和使用方法，將其作為創作過程中的一部分。

結合傳統技術：將 AI 技術與傳統創作技術結合，充分利用 AI 的自動化和創意輔助功能，提高創作效率和作品質量。

保持創意

強化個人風格：在使用 AI 生成影像時，創作者應該注重保持和強化自己的個人風格，避免作品過於同質化。

探索新表現形式：利用 AI 技術探索新的藝術表現形式和創作方法，不斷突破傳統創作的限制，開創更多元化的藝術表達。

倫理和責任

尊重版權：在使用 AI 生成影像時，應該注意尊重他人的版權和知識產權，避免侵犯到他人的創作權利。

負責任使用：創作者應該負責任地使用 AI 技術，避免生成不當或具爭議性的內容，維護創作的道德標準。

總而言之，AI 生成影像技術為攝影和繪畫藝術創作帶來了新的機遇和挑戰。創作者需要調整心態、擁抱技術、保持創意，並負責任地使用 AI 工具，以在這個不斷變化的藝術世界中保持競爭力。

開啟全民創作時代

生成式 AI 技術的發展，降低了創作門檻，使得任何人都能輕鬆創作高質量的圖像、文字、音樂等藝術作品，而無需專業技能或設備。這些 AI 工具，如 Midjourney、DALL.E 和 Adobe Firefly，只需簡單的文本描述就能生成精美的作品，顯著提高了創作效率。AI 不僅能夠快速生成作品，還能提供創意輔助，幫助創作者探索新的藝術表現形式。此外，AI 生成技術還促進了個性化創作，使

得用戶可以根據自己的喜好和需求，創作出高度定制化的作品。

這些技術同時也促進了社交分享和合作創作，讓創作者能夠輕鬆分享作品並與他人交流經驗。總之，AI技術的進步推動了創作的民主化和多樣化，使得每個人都能發揮無限創意，成為獨特的創作者。

但無論是傳統的創作者或是即將透過AI工具進行創作的人，我們都必須理解在所謂AI時代，「技巧」已不是最關鍵的因素，更多的是觀念，是我們長期忽略的人文素養，包括對歷史、藝術、美學、生活觀察等多方面的理解和敏感性。

● 歷史知識

基於團隊模型的標示，理解歷史能夠讓創作者在AI生成影像時賦予作品更多的背景和深度。例如，對某一時代的建築風格、服飾、藝術風格的了解，能夠讓AI生成的影像更符合歷史真實性和美學要求。這種知識可以幫助創作者提供更精確的提示詞，從而生成更具說服力和文化內涵的影像。

● 生活觀察

生活觀察能力是指對日常生活細節的敏銳感知和記錄。這種能力能夠幫助創作者捕捉到細微的情感變化、場景氛圍和人物特徵，從而在提示詞中包含更多具體而生動的描述。例如，描述「黃昏時分，老人坐在公園長椅上閱讀」這樣的場景時，細節的豐富性來自於對生活的深入觀察。

● 藝術修養

具備一定的藝術修養，了解不同的藝術流派、風格和技法，能夠讓創作者在使用AI時提供更具體的藝術指導。例如，要求生成一幅梵高風格的夜景畫時，理解梵高的筆觸和色彩運用可以提供更具體的提示詞，如「使用粗重的筆觸和鮮亮的色彩，模仿梵高的《星夜》風格」。

● 文學素養

文學素養體現在對文字表達的敏感性和理解力上。高水平的文字表達能力可以使提示詞更加生動具象，讓AI生成的影像更具感染力。例如，描述一個浪漫的場景時，可以使用更具詩意的語言，如「夕陽染紅了天空，戀人們在沙灘上牽手漫步」。

在實際創作中，這些人文素養往往需要綜合應用。例如，創作一幅反映十九世紀工業革命時期城市景觀的圖像，創作者需要運用歷史知識來設定場景，用生活觀察能力來描述細節，用藝術修養來決定圖像的風格和構圖，用文學素養來撰寫生動的提示詞。

擁有人文素養能夠讓創作者在使用生成式AI進行創作時，更加自信且具有創造力。這些素養不僅豐富了創作的內涵，更能提升作品的質量、獨特性和影響力。隨著AI技術的不斷進步，AI勢必取代多數技能性的工作，而能夠驅動AI的，必是人文訓練。

初學者如何進入 AI 生成

「工欲善其事，必先利其器」，在學習和使用生成式 AI 影像工具時，理解和掌握這些工具的特性和操作是至關重要的。

在眾多的生成式 AI 影像工具中，我們並非獨厚 Midjourney，但就學習進入門檻及生成結果而論，它的確是目前最能滿足所有人的工具平台。因此我們也多花了一些篇幅介紹 Midjourney 的命令及參數，並且特別增加 PART4 單元，專門提供給本書讀者的 Midjourney 專屬風格用法。

本書並不鎖定為 Midjourney 的使用指南，但相信你透過這本書的學習與了解，一定可以將 Midjourney 強大的功能發揮至淋漓盡致，透過本書一步步引導的提示詞的寫法，可適用在 OpenAI DALL.E、Google Gemini、Adobe Firefly 等生成工具，也同樣通用在多數的影音生成平台，這些提示詞的寫法都能夠幫助讀者在這些生成平台上取得良好的結果。

除了透過本書學習外，建議讀者們多方練習與加入線上討論，讓 AI 生成功力更精進。一起加入吧！

▶ 安裝 Discord

使用 Midjourney 有二種方式，第一種是最多人早期使用的 Discord，這是 Midjourney 主要的使用介面之一。第二種是後來 Midjourney 才提供的 Web 介面。Discord 是由美國 Discord 公司所開發的一款專為社群設計的免費網路即時通話軟體與數位發行平台，主要針對遊戲玩家、教育人士、朋友及商業人士，使用者之間可以在軟體的聊天頻道透過訊息、圖片、影片和音訊進行交流。下載位置：https://discord.com/download

可以在這地方找到各種支援的系統，包括 Android 手機、iPhone 手機，及 Windows、iOS、Linux 等不同的電腦操作系統。安裝完後，可以在 https://discord.com 的網頁上了解一下基本的操作，原則上介面跟瀏覽器及社群軟體很接近，礙於篇幅，我們無法在這裡做詳盡的介紹。

安裝完成後，可以**掃描下面 QRcode 加入「Louis Huang 的教學伺服器」**來進行進一步的學習和交流。本書的一些勘誤，或是讀者的問題，都可以在上面提出，我會一一回答。更重要的是，這個地方將成為我們共同研討、共同創作的專屬伺服器。

Discord 下載

Discord 網站

Louis Huang 的
教學伺服器

人人都會 AI 繪圖
FB 社團

▶ 訂閱主要工具平台

本書中提到的主要生成工具平台如 Midjourney、Copilot、Gemini 和 ChatGPT 均為付費平台，部分提供試用或免費使用額度。這些平台目前對繁體中文的支援程度不一，建議讀者在使用這些工具時，優先考慮以英文建立提示詞，以獲得更理想的生成結果。對於不熟悉英文的讀者，可以利用各類免費翻譯工具來輔助撰寫提示詞。※ 具體的需求及說明以各自官網為主。

https://www.midjourney.com
https://www.microsoft.com/zh-tw/microsoft-copilot
https://gemini.google.com
https://chatgpt.com
https: //firefly.adobe.com

在完成安裝及基本訂閱後，我們便可以開始嘗試用提示詞生成影像，在這之前還有個關鍵就是語言，正常來說，這些平台均宣稱有支援繁體中文，但截至目前，Midjourney、Gemini 的語言支援並不理想，而微軟及 Open AI 的中文支援程度較高。另外專業工作者常使用 Stable Diffusion 目前也不支援中文提示詞。此外，我們必須知道這些 AI 工具，原生的模型訓練及 token 的建立、對應均是以英文進行，所謂中文支援是將提示詞再進行一次翻譯去對照英文 token，因此生成結果或許也會有期待落差，最好的做法當然還是以建立英文提示詞為最優先。

這對母語非英語系使用者而言，是有困難的，但所幸目前有許多翻譯工具，包括 google, Microsoft 等等，都有提供免費的線上翻譯服務，甚至文書處理的 Office Word 也有內建翻譯功能。相信在這些工具的協助之下，以英文建立提示詞將不是太大的困難。

本書每則範例均會附上英文原文的提示詞，讀者也可以參照，如需直接引用，除了直接上網下載外，也可手動輸入或透過手機軟體，將文字拍攝下來，再進行文字選取功能，複製貼上即可。

| Midjourney | Copilot | Gemini | Chatgpt | Adobe Firefly |

▶ 關於智慧財產權

隨著生成式 AI 技術的快速發展，AI 生成影像在智慧財產權（IP）方面引發了諸多爭議和討論。這些爭議主要集中在以下幾個方面：

◆ 版權歸屬問題

生成式 AI 影像的版權歸屬問題是目前最受關注的爭議點之一。在傳統上，版權通常屬於創作者，即創作內容的人。然而，當影像是由 AI 生成時，創作者的定義變得模糊。

AI 工具的開發者：部分觀點認為，AI 工具的開發者或擁有者應該擁有版權，因為他們提供了生成這些影像的技術平台和工具。

AI 用戶：另一部分觀點則認為，使用 AI 工具生成影像的用戶應該擁有版權，因為他們提供了創意、提示詞以及生成的方向和指導。

公共領域：還有觀點認為，AI 生成的影像應該進入公共領域，因為這些作品並非由人類直接創作，而是經由算法產生。

◆ 訓練數據的版權問題

AI 模型的訓練數據來源也是一個重要的爭議點。AI 生成影像的質量和風格很大程度上依賴於其訓練數據，而這些數據通常包含大量受版權保護的圖片。

未經授權使用：如果 AI 模型在未經版權持有者授權的情況下使用了受保護的圖片進行訓練，這可能構成版權侵權，進一步引發法律糾紛。

授權和補償：一些平台開始探索如何合法使用受保護的內容進行訓練，包括獲取授權並向版權持有者提供補償。例如，Adobe Firefly 在其訓練數據中使用了授權的 Adobe Stock 圖片【79†source】，試圖以合法途徑來解決這一問題。

◆ 衍生作品的認定

生成式 AI 影像是否屬於原始訓練數據的衍生作品，這一點在法律上尚無明確結論。

衍生作品：如果 AI 生成的影像被認定為原始訓練數據的衍生作品，那麼這些影像的使用可能需要得到原始版權持有者的許可，否則會被視為侵犯版權。

新創作：如果 AI 生成的影像被視為全新的創作，那麼這些影像將擁有獨立的版權，不需要原始版權持有者的許可。但這樣的認定在實際操作中可能會遇到法律上的挑戰。

◆ 道德和倫理考量

除了法律問題，AI 生成影像的智慧財產權還涉及道德和倫理問題。

創作者權益：如何保護人類創作者的權益，防止他們的作品被 AI 未經授權地使用，是一個重要的倫理問題。這也涉及到如何公平地對待和補償原始創作者。

透明度：AI 生成影像的過程應該是透明的，讓使用者和公眾了解這些影像是如何生成的及訓練數據的來源，以確保這一過程符合倫理標準。

這些爭議表明，隨著生成式 AI 技術的發展，智慧財產權法律和倫理規範需要與時俱進，以適應新技術帶來的挑戰。創作者、技術開發者和法律制定者需要共同努力，在原始藝術創作者、工具平台提供者和 AI 使用者之間尋求平衡，確保 AI 技術的發展既促進創新，又保護智慧財產權和創作者的權益。

目前，智慧財產權領域仍處於快速發展和變化之中，我們期望未來能有更完善的法律和規範出現，能夠在原始藝術創作者、工具平台提供者、AI 使用者三者間達到一個平衡，而不是偏廢。

整體而言，只要不是以任何形式將非屬於自己的生成標的「複製」使用，便無侵權疑慮。這意味著，只要是你自己輸入提示詞，再由 AI 生成的物件，是沒有侵權的問題，相反地你還擁有該生成物主張的權利。但這權利基本也不受保護，因為在一定機率下，也會有其他的使用者用相同的提示詞生成相同的圖片。

> 以本書為例，本書所生成的影像是屬於作者及出版單位的著作權保護範疇，任何人不能直接進行複製或轉載。但本書所提供的提示詞，讀者可以自行使用，因為提示詞在實務角度很難形成著作權保護。讀者輸入後，AI 工具大概率會生成類似但不完全相同的影像，在這種情況下，原作者亦不能主張其著作權。
>
> 這反映出 AI 生成影像領域的法律複雜性與挑戰，也進一步強調了在使用 AI 工具時的謹慎和自我保護意識。

DATA DOWNLOAD

https://reurl.
cc/6jbZVO

全書「中英文提示詞」檔案下載

本書特別為購書讀者準備「中英提示詞」，軟體為 Excel，均可至下列網址下載，請各位盡情運用在自己的創作中。

※ 嚴禁上傳到社群網站、在任何載體發布、從圖書館借閱禁止使用此檔。

如何使用

● 書中標誌 🖥 代表有可以下載的「中英提示詞」。
● 下載檔名皆以書中頁碼編號，方便查詢。

下載圖示 ── 🖥 **072** ── 頁碼

參考資料

- [Adobe Firefly - Free Generative AI for creatives](https://www.adobe.com/products/firefly.html)【79†source】
- [OpenAI on DALL.E 2: Generative Art and Copyright Issues](https://www.openai.com/dall-e-2)

PART 2

關鍵提示詞及
深度使用
Midjourney

一、基本提示詞

在 Midjourney 平台上，Midjourney 機器人（Midjourney Bot）會將我們輸入的提示詞分解成更小的部分，稱為 Token，這些 Token 會與其訓練資料進行比較，並用來生成圖像。這種處理方式也是目前所有文生圖工具平台的共通邏輯，提示詞與訓練資料（圖像模型）匹配得越精準，生成的影像就越接近我們的期望。

一個基本的提示詞可以是單詞、短語，甚至是表情符號。即使提示詞是無意義的，AI 平台仍能隨機生成影像。因此，使用 Midjourney 機器人時，建議以簡單的短語來描述你想要的內容，避免使用冗長的請求或說明。

■　以下部分圖文引用自 Midjourney 官網，www.midjourney.com。

範例

提示詞 ⇒ **Show me** a photo of blooming Prunus subgen, **make them** bright, vibrant orange, and **draw them** with colored pencils.

給我看一張盛開櫻花的照片，讓它們明亮、鮮豔的橙色，並用彩色鉛筆畫出來。

提示詞 ⇒ Bright orange Prunus subgen drawn with colored pencils.

用彩色鉛筆畫出的明亮橙色的櫻花。

✕ 錯誤提示

○ 正確提示

結論

比較上面二張圖，基本上是相同的，也就是說，**沒有意義的詞彙**在提示詞中只是多餘，不會影響生成結果。目前在網路上流傳一些很奇妙的「咒語」，事實上 AI 生成結果並不會被那些情緒性、甚至語助詞所影響。

■　小提示：利用 Midjourney /shorten 命令，可以快速精簡提示詞中的贅詞。

這樣會更好！

此外，提示詞也該避免使用主觀的形容詞，例如酷（Cool），如果我們做個「Cool 的形象」問卷調查，可能會得到數十種答案，當然你也可以一直用這個字當「咒語」直到 AI 生成出你期待的影像。例如：

提示詞 ⇒ A cool young man.
一個酷的年輕人。

不是說這種作法不好，如果當下沒有甚麼主見，用這種空泛的形容詞，也許可以透過 AI 找到一些靈感。提示詞標示越精準明確，也就對 Cool 有清楚的定義，例如：

提示詞 ⇒ A young man with purple-dyed hair, a tiger tattoo, wearing riveted pants, sunglasses, and a baseball cap.

一個年輕男人，染紫色的頭髮，身上有老虎的刺青，穿著有鉚釘的褲子，戴墨鏡，棒球帽。

二、進階提示詞

進階提示詞可以包含一個或多個圖像 URL、多個文字短語和參數設定。這也是本書接下來將介紹的內容。

● **Image Prompts（圖像提示詞）**：你可以直接提供圖像的連結網址，將其加入提示詞中，以影響最終結果的風格和內容。「圖像的連結網址」要放在提示詞的最前面。

● **Text Prompts（文字提示詞）**：這是你用來描述希望生成的圖像的文字。精心撰寫的提示可以幫助產生更為出色的圖像效果。請參閱下面的提示資訊和提示。

● **Parameters（參數提示詞）**：參數可以改變圖像生成的方式。參數可以改變長寬比、模型選擇、變更圖像的風格或形態、進行人物、風格參照或增強變異度等等。參數通常放在提示詞的末端。

▶ 提示詞的撰寫技巧

◆ 詞語選擇

詞語選擇很重要。在許多情況下，更具體的同義詞通常能帶來更好的效果。例如：不要使用 big，可以試試 tiny, huge, gigantic, enormous 或 immense。

◆ 複數詞和集合名詞

複數詞可能帶來很多不確定性。請嘗試使用具體的數量。例如：「三隻貓」比「貓」更具體。集合名詞也有效，例如：使用「鳥群」而不是「鳥」。

◆ 聚焦你想要的內容

描述你希望看到的，而非你不希望看到的。例如：如果你要求一個「沒有蛋糕」的派對，你的圖像可能會包含蛋糕。要確保某個物體不出現在最終圖像中，可以使用參數 --no

◆ 提示詞長度和細節

提示詞可以很簡單，如一個單詞或表情符號。但是，簡短的提示詞會依賴 Midjourney 的默認風格，自行填補未指定的細節。因此，提示詞中非常重要的元素就是細節關鍵。**描述細節越少，變化越多，控制力越小**。對於創作者而言，我們應該盡力去控制畫面生成的要素，以期達到期望的影像。撰寫提示詞時，務必清楚描述重要的背景或細節。

● 提示詞撰寫方向參考

主題	人物、動物、角色、地點、物體。
媒材	照片、繪畫、插圖、雕塑、塗鴉、掛毯。
環境	室內、室外、月球上、水下、城市中。
光照	柔和、環境光、陰天、霓虹燈、攝影棚燈光。
顏色	鮮豔、柔和、明亮、單色、多彩、黑白、粉彩。
氛圍	沉靜、平和、喧鬧、充滿活力。
構圖	肖像、頭部特寫、特寫、鳥瞰圖。

我們可以從一個空白畫面開始模擬「如何產生一個充滿細節」的提示詞：

❶ 一隻貓。

❷ 一隻波斯貓。

❸ 一隻藍色波斯貓。

❹ 一隻表情憤怒的藍色波斯貓。

❺ 一隻表情憤怒跳起來的藍色波斯貓。

❻ 水彩畫，一隻表情憤怒跳起來的藍色波斯貓，在公園……。

❼ 水彩畫，一隻表情憤怒跳起來

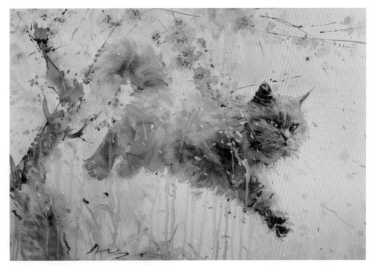

這是以 7 為提示詞所產生的 AI 生成圖。

的藍色波斯貓，在公園，櫻花樹，春天的早晨，下雨天，朦朧的氛圍，Andrew Nowell Wyeth 的繪畫風格。

> A watercolor painting of an angry blue Persian cat leaping into the air in a park. The scene is set under a cherry blossom tree on a rainy spring morning. The delicate pink blossoms are in full bloom, with raindrops gently falling and creating a soft, misty atmosphere. In the style of Andrew Wyeth --ar 3:2

◆ 清晰的邏輯才有成功的影像

撰寫提示詞是一個逐步構建畫面的過程，從無到有，從簡單到複雜，從粗略到細節。無論是圖像、文字、還是音樂的創作，基本上都遵循這樣的邏輯。但**提示詞不是作文，不需要文詞優美，但必須清晰有內容、言之要有「物」**。在生成式 AI 影像平台剛問世時，甚至現在，仍有許多人不懂 AI 影像生成的底層邏輯，便將提示詞說成是「咒語」，但現實是「天靈靈地靈靈」或「芝麻開門」不會生成你想要的畫面，只是有時或許運氣好，AI 隨機生成了一個你覺得還可以的畫面。成功的影像生成提示詞，不靠人品或運氣，而是清晰的邏輯。

撰寫有效提示詞需要深思熟慮和精確描述。首先，需要明確想要的主題和氛圍。例如，如果想生成一個溫暖的家庭場景，可以使用像「客廳，柔和燈光，家人在一起聊天」這樣的提示詞。這樣的描述明確且具體，能夠幫助 AI 理解關鍵元素並生成符合預期的圖像。

接著，提示詞應該包括具體的細節來增強圖像的準確性。例如，想要生成一幅城市夜景，可以添加更多細節，例如：城市街道、霓虹燈閃爍、人群、車輛、夜空中有星星……等。這些細節有助於 AI 理解並構建更精細和豐富的畫面。

最後，提示詞應描述構圖、光線、色調等視覺元素，以進一步提升生成影像的質量。例如：想要一幅浪漫的沙灘日落圖像。你可以這樣描述：金黃色的沙灘，波光粼粼的海面，橙紅色的夕陽，兩人在海邊牽手散步。這樣的提示詞不僅描繪了場景，還突顯了情感細節和氛圍。

▶ 提示詞在不同 AI 工具的效果

撰寫提示詞是一個反覆推敲和完善的過程。通過精確描述和詳細設定，使用者可以更好地控制 AI 生成的影像，從而創造出更符合自己期望的作品。理解並掌握提示詞的撰寫技巧，能夠顯著提升 AI 影像生成的效果和創作滿意度。

本書主要是以 Midjourney 作為生成工具，會詳細介紹 Midjourney 所需要的各種命令及參數。不過，即使忽略這些專用命令及參數語法，書中所介紹的提示詞撰寫邏輯及技巧，只要是透過自然語言模型作為驅動的 AI 工具，便可適用在所有 AI 影像生成工具。例如：Microsoft Designer、Copilot、DALL.E、Google Gemini、Adobe Firefly 和 Stable Diffusion 等生成工具，亦可使用於一些動畫生成工具，例如：SORA, Runway, Leonardo 等。

以下範例展示了在不同平台上，使用相同提示詞（刪除 Midjourney 專用參數後）所生成的結果：

> A watercolor painting of an angry blue Persian cat leaping into the air in a park. The scene is set under a cherry blossom tree on a rainy spring morning. The delicate pink blossoms are in full bloom, with raindrops gently falling and creating a soft, misty atmosphere. In the style of Andrew Wyeth --ar 3:2
>
> 水彩畫，一隻表情憤怒跳起來的藍色波斯貓，在公園，櫻花樹，春天的早晨，下雨天，朦朧的氛圍，Andrew Nowell Wyeth 的繪畫風格。

1.Microsoft Designer ／ Copilot ／ DALL.E

這三個工具使用路徑及介面有部分差異，但影像生成的底層模型及演算法都是使用 OpenAI 的 DALL.E。

2. Google Gemini

3. Adobe Firefly

每個工具平台或略有差異，但以文本生成影像的邏輯是相通的。上面的範例，每個人或許都會有偏好，但很明顯地只有 Midjourney 可以表現出指定的藝術風格，而其他或許因為模型訓練的標的差異，僅能出現媒材（水彩畫）的視覺效果，但主題的完成度相當一致。

● 認識 AI 圖像產生器

生成器	特色	使用介面	價格	發行公司
Midjourney	藝術風格。	Discord、Web 網頁及應用程式	200 張圖像，每月 10 美元起。	Midjourney
DALLE·3	最容易上手的生成器。	ChatGPT、Bing、Copilot 網頁及應用程式。	1. 免費會員：每天 2 張免費圖片。 2.ChatGPT Plus： 每月 20 美元起。	OpenAI
Ideogram	準確的文字與 AI 影像結合的能力。	Web 應用程式。	1. 有限的免費計劃。 2. 每月 8 美元起，可高解析度下載和 400 使用積分。	Ideogram AI
Stable Diffusion	可客制完整的生成功能，包括影像、影片、3D 等等。	NightCafe、Tensor.Art、Civitai 以上應用程式都是以 SD 為核心，搭配不同的使用者介面。	1. 第三方平臺自行收費。 2. 開源模型免費，需自行架設伺服器。 3. 官方 Stable Artisan & Assistant，每月 9 美元起。	Stability AI
Adobe Firefly	AI 增生。	firefly.adobe.com、Photoshop、Express 和其他 Adobe 工具。	每月台幣 156 起。	Adobe
Generative AI by Getty	可安全商業化的圖像。	iStock 網頁。	每月 24.99 美元起。	Getty（uses NVIDIA Picasso）

在語言的支援上，大多都有支援中文，但我們還是建議提示詞均以英文書寫，這是基於平台工具的開發邏輯，用英文提示詞生成的影像，可以降低因為語言轉換所造成的期待落差，畢竟理解語言不是 AI 影像生成工具的訴求。

▶ 提示詞生成四要素

「提示詞」就是生成式 AI 影像最關鍵之處，簡單來說，誰的提示詞寫得好，誰的圖像就生成得好。然而，不同於寫作中的起承轉合，提示詞的撰寫更注重的是精確的表達和清晰的結構。除了特定平台工具所需的命令跟參數外，我們將提示詞分為四大部分：媒材、主題、構成和風格。其中媒材、構成、風格會在 PART3 及 PART4 單元做完整的介紹。

關於主題，就是對人、事、時、地、物的描述，也就是寫景、敘事、描物的過程。這個邏輯不難理解，但要寫得「言之有物」，則需要平時對事物的細心觀察。Midjourney 的創辦人 David Holz 甚至直言：「使用 Midjourney 不僅僅是學習如何使用這個工具，更是學習所有的藝術和歷史」。

● 提示詞生成四要素

媒材					
主題	人	事	時	地	物
構成	視角	顏色	燈光	表情	場景	氛圍
風格					

三、命令列表説明

在 Midjourney 平台上，使用「/」命令（commands）來與 Midjourney Bot 進行互動並控制圖像生成的各種參數。這些命令可以用來調整生成的圖像、修改參數或執行特定的功能。以下是一些常用的 Midjourney「/」命令及其作用：

Midjourney 命令可以在任何機器人頻道、允許 Midjourney 機器人運行的私人 Discord 伺服器上，或與 Midjourney 機器人的直接訊息中使用。以下我們列出較重要及常使用的命令，更多的命令及更新，請隨時參考官方説明。 https://docs.midjourney.com/docs/command-list。

▶ 基本指令

1. /blend（混合）

輕鬆將兩個圖像混合在一起。輸入指令後，系統會提示您上傳兩張照片。您也可以選擇更多影像進行混合，並調整長寬比。

> 使用行動裝置時，從硬碟拖曳影像或從照片庫新增影像。若要新增更多影像，請選擇該 optional/options 欄位並選擇 image3、image4 或 image5。該 /blend 命令可能比其他命令需要更長的時間來啟動，因為必須先上傳您的圖像，然後 Midjourney 機器人才能處理您的請求。
>
> 混合影像的預設長寬比為 1:1，但您可以使用選用欄位 dimensions 在方形長寬比（1：1）、縱向長寬比（2：3）或橫向長寬比（3：2）之間進行選擇。
>
> /blend 與任何其他提示一樣，自訂後綴會新增到提示的末端 /imagine。作為命令一部分指定的長寬比 /blend 會覆蓋自訂後綴中的長寬比。

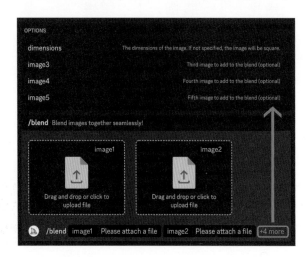

2. /describe（描述）

以圖生文，根據您上傳的圖像編寫四個範例提示。這是非常好用的功能，當看到一幅影像但卻不知道該如何下提示詞時，使用 /describe 命令，便可以得到一些屬於該圖像的詞彙和主題風格提示。但請注意，這個功能的目的是產生啟發性和暗示性的提示，不能用於準確地重新建立上傳的圖像。

3. /imagine（生成影像）

使用提示產生圖像的命令。這個命令應該是最重要也最常用的！在訊息欄位中輸入「/imagine」。您也可以 /imagine 從鍵入 " / " 時彈出的可用斜杠命令清單中選擇命令。在欄位元中輸入要建立的影像的描述 prompt。

4. / prefer remix（開啟或關閉混合模式）

使用命令 /prefer remix 或使用 /settings 命令並點擊 ♻ Remix Mode 按鈕來開啟混合模式。混合模式啟動後，重新混合模式可讓您在進行變體（包括 Vary（Region）)、重新生成 🎲 或平移現有影像時再次編輯提示詞。

使用重新混合模式時，除了提示文字之外，您還可以編輯影像參考、模型版本、參數和權重，但產生的提示元素組合必須仍然有效。如果您不對 Remix 視窗中的提示進行任何更改，Midjourney 將產生正常變體。

操作解說：

❶ Remix Mode 啟用後，V 點擊按鈕時按鈕會變成綠色而不是藍色。
❷ 出現 remix 提示視窗，再輸入提示詞，這個範例我們希望原圖天空中出現幽浮。
❸ 完成。

5. /prefer suffix（偏好參數設定）

指定要新增到每個提示末尾的參數。

命令範例：/prefer suffix --ar 16:9 –video

若要清除設定的參數，請使用該 /settings 命令並選擇 Reset Settings，或 /prefer suffix 再次使用而不向命令添加可選的 new_value 欄位。

這是個很實用的小功能，如果創作者已經非常確定或者常使用的各項參數設定值，便可使用此功能將部分參數設為預設值，免去重複輸入的麻煩。

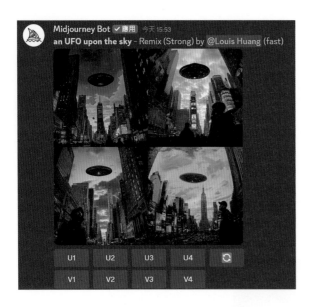

6. /prefer variability（偏好變化設定值）

Midjourney 的變化設定值有二種模式，預設是 High variability 高度變化，也就是在生成圖片後使用下方的 V1, V2, V3, V4 按鈕，重新生成圖像的變化強弱。鍵入 /prefer variability 可以切換高、低變化值，或者可以從 /setting 中去調整預設值。

7. /relax /fast /turbo（生成速度）

切換到放鬆模式，或稱為一般模式。不同層級的訂閱者享有不同時數的加速模式，使用者可以視作業需求程度在 /turbo /fast /relax 三者間切換。

8. /setting（設定）

查看並調整 Midjourney 機器人的設定。

■ 新增至提示末端的參數將覆蓋使用 /settings 所做的選擇。

9. /stealth, public（隱形或公開模式）

專業計劃訂閱者可切換到隱形模式。

Midjourney 是一個預設開放的社群平臺，所有圖像生成都可以在 midjourney.com 上看到，包括在私人 Discord 伺服器和 Midjourney Web 應用程式上創建的圖像。

但如果付費加入 Pro（月費 60 美金）和 Mega（月費 12 美金）計劃的訂閱者可以使用隱形模式。隱形模式可防止 Midjourney 網站上的其他人看到您的影像。使用 /stealth 和 /public 命令在隱密模式和公共模式之間切換。

隱形模式只會阻止其他人在 Midjourney.com 上查看您的圖像！

即使在使用隱形模式時，公共管道中產生的圖像始終對其他使用者可見。為了防止其他人看到您使用隱形模式建立的影像，請在私人頻道或私人 Discord 伺服器上產生影像。

10. /shorten（縮短）

提交長提示詞並接收有關如何使其更簡潔的建議。

這個功能也是十分實用的功能，有時候我們可以使用一篇文章作為提示詞，再透過 /shorten 將文章變成具有意義的影像生成提示詞。從中再學習 AI 對於提示詞的理解與識別度，進而提高日後撰寫提示詞的準確性。

11. /show（展示）

使用圖像 job ID 在 Discord 中重新產生作業。

如何取得 job id：在 midjourney.com 上點選任何自己所生成的圖像，然後找到漢堡選單 → Copy → Job ID，便可將 Job ID 複製在剪貼簿上。

對於一些重度或專業工作者而言，生成圖片數量龐大，可以用這功能呼叫出當初下的提示詞，或者對原圖進行放大或其他變化調整。

12. /synonyms（同義詞探索）

使用者可以在「官方 Midjourney Discord 伺服器」鍵入 /synonyms（words）來探索相關單字或片語以在日後的提示詞中嘗試。現在的 AI 平臺是一個會自我進化的系統，即便是最初的研發工程師也很難說得清楚當初開發的 AI 平臺，現在進化到甚麼程度。我們可以使用 /synonyms 同義詞探索這功能，找出更多有意義且精準的提示詞。

四、參數列表説明

參數是添加到提示中的選項，用於調整圖像生成的方式。例如：您可以改變圖像的長寬比、切換不同的 Midjourney 模型版本，或選擇使用的放大器（Upscaler）。參數通常放在提示的末尾，且每個提示可以包含多個參數。更多完整及更新的參數使用請隨時參考官方説明。 https://docs.midjourney.com/docs/parameter-list。

 /imagine `prompt` `California poppies --aspect 2:3 --no sky`

	參數	説明
1	--ar <X:Y>	功能：調整生成圖片的長寬比例。 説明：使用 --ar 16:9，可以創建寬屏圖像。 這是 Midjourney 上非常重要的功能，其他的平台只能做橫幅或直幅的調整。而通過長寬比例調整，便可以改變構圖，例如要生成全景影像，就要設到 3:1 甚至更寬。目前也只有 midjourney 可以精準控制畫面比例，如 23:9, 17:5 等。
2	--chaos <0–100>	功能：調整生成圖片的隨機性。 説明：數值越高，圖像會越不尋常、越有創意。使用 --chaos 75，可能會產生更多意想不到的結果。
3	--cref <image url>	功能：使用指定圖片作為角色參考。 説明：可幫助生成多張圖片時保持角色一致性。
4	--fast	功能：強制系統使用快速模式完成單一作業。 説明：適合希望儘快完成生成的情況。
5	--iw <0–3>	功能：調整提示詞與圖像的權重。 説明：預設值為 1，數值越高，原始圖像對生成結果的影響越大。使用 --iw 2.5，會更偏向於原始圖像的風格。 ◎請見【範例 01】
6	--no <object>	作用：排除物件、負面提示。 説明：使用 --no plants，可以排除圖片中的植物。
7	--relax	作用：強制系統使用放鬆模式完成單一作業。 説明：適合時間不趕時使用。
8	--repeat <1–40> 或 --r <1–40>	作用：重複生成多張圖片。 説明：使用 --repeat5，會基於同一提示生成 5 張圖片。
9	--seed <integer between 0–4294967295>	作用：指定生成圖像時的隨機種子編號。 説明：使用相同的種子和提示詞會產生相似的圖像。 ◎請見【範例 02】

10	--stop <integer between 10–100>	作用：在生成過程中提前停止。 說明：可以控制生成的細節程度，停止越早，圖片越模糊。
11	--style <raw>	作用：使用原始風格生成圖片。 說明：適合已經熟悉提示詞且希望更多控制生成結果的用戶。 ◎請見【範例 03】
12	--sref <image url>	作用：指定樣式參考圖片。 說明：生成圖片時會參考該圖片的風格或美感。 ◎請見【範例 04】
13	–sw <0-1000>	作用：調整樣式化的強度。 說明：數值越高，生成圖片會更符合設定的風格。 ◎請見【範例 05】
14	–sref random	作用：隨機指定風格參考。 說明：系統會隨機生成一個風格參考值，效果會有所不同。但目前這串數字並沒有規律定義，只知道總數值數等於種子數，也就是高達 43 億！ ◎請見【範例 06】
15	--stylize <number> 或 --s <number>	作用：調整系統預設的美學風格強度，可以產生有利於藝術色彩、構圖和形式的影像。 說明：數值越低，生成圖片越貼近提示詞；數值越高，生成圖片會更有藝術感，但與提示詞的關聯度降低。 stylize 的預設值為 100，使用目前模型時接受 0–1000 的整數值，我們直接引用官網上的範例來檢視一下不同數值與提示詞之間的關係。 ◎請見【範例 07】
16	-tile 連續圖案	作用：生成可重複排列的連續圖案。 說明：適合用來創建連續圖案的背景圖片或四方連續圖案的圖像。 ◎請見【範例 08】
17	–turbo	作用：使用加速模式完成單一作業。 說明：比快速模式更快，但可能消耗更多資源。
18	--weird <number 0–3000> 或 --w <number 0–3000>	作用：調整圖片生成的異常程度。 說明：數值越高，圖片風格越異常，適合探索不同風格的美學。

19	預設值（模型版本 6）		縱橫比	混亂	品質	種子	停止	風格化
		預設值	1:1	0	1	隨機的	100	100
		範圍	1:14 – 14:1	0 – 100	.25 .5 或 1	整數0 – 4294967295	10 – 100	0 – 1000

20	--version <1, 2, 3, 4, 5.0, 5.1, 5.2, 6 or 6.1> 或 --v <1, 2, 3, 4, 5.0, 5.1, 5.2, 6 or 6.1>	作用：指定 Midjourney 使用的模型版本。 說明：Midjourney 會定期發布新版本的模型，每個版本都針對不同的影像類型進行了優化和改進。選擇適合的版本可以幫助你生成更符合預期的圖像。例如，--v 6 使用最新的第 6 版模型。
21	–niji <4, 5, 6>	作用：切換到指定版本的動漫風格模型。 說明：這個參數適用於創建帶有動漫風格的圖片。選擇不同版本（例如 --niji 5）可以生成不同風格的動漫圖像。

許多 Apple 裝置會自動將雙連字號（--）變更為長破折號（─），但 Midjourney 接受兩者！

範例 01 | --iw <0-3>

/imagine prompt flowers.jpg birthday cake –iw <.25-3>

可以很明顯的看出在權重 .25 時是以提示詞蛋糕為主，到了權重為 3 時，則是以原始圖片為主。

原始圖片　　　--iw .25　　　--iw .25　　　--iw .75

--iw 1　　　--iw 1.25　　　--iw 1.5　　　--iw 1.75

--iw 2　　　--iw 3

範例 02 | --seed <integer between 0—4294967295>

如何取得 seed number?

在官網的個人網頁上,選取圖片→選取選單→ Copy → Seed,便可將 Seed Number 複製在剪貼簿。

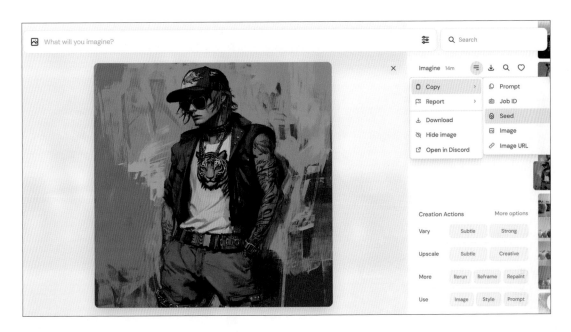

使用 Seed 來控制風格統一性是非常好用的,例如上圖,我們將年輕男子改成年輕女性,使用相同 **seed number 1651839283**。

提示詞 ⇒ A young woman with purple-dyed hair, a tiger tattoo, wearing riveted pants, sunglasses, and a baseball cap --seed 1651839283

--style raw 原始風格，對於已經習慣完善提示詞並希望對其圖像進行更多控制的用戶來說，原始風格更能貼近提示詞的描述。--style raw 應用了較少的自動美化，這可以在提示特定樣式時實現更準確的匹配。（本範例引用自 Midjourney.com 官網）

--v 6　　　　　　　　　　　　　　　　**--v 6 --style raw**

提示詞 ⇒ black and white oak tree icon

提示詞 ⇒ black and white oak tree icon **--style raw**

--v 6　　　　　　　　　　　　　　　　**--v 6 --style raw**

提示詞 ⇒ hild's crayon drawing of a puppy

提示詞 ⇒ child's crayon drawing of a puppy **--style raw**

在 Midjourney 模型版本 5.1 和 5.2 之間切換。在 **--style <4a, 4b, or 4c>**Midjourney 模型版本 4
的版本之間切換。

--style <cute, expressive, original, or scenic>

範例 04 | **--sref <image url>**

將圖片網址 **https://s.mj.run/P2RgB6onCCg** 作為風格參照。

風格參照原圖

提示詞 ⇒ a dog --sref **https://s.mj.run/
P2RgB6onCCg**

範例 05 | **-sw <0-1000>**

接續【範例 04】的例子，使用不同的風格 sw。

提示詞 ⇒ a dog –sref https://s.mj.run/
P2RgB6onCCg **–sw 50**

提示詞 ⇒ a dog –sref https://s.mj.run/
P2RgB6onCCg **–sw 800**

範例 06 │ -sref random

提示詞 ⇒ A dog –sref random（系統隨機生成數值 3572099357）

提示詞 ⇒ a dog –sref random（系統隨機生成數值 3574895463）

提示詞 ⇒ a dog –sref random（系統隨機生成數值 3529758812）

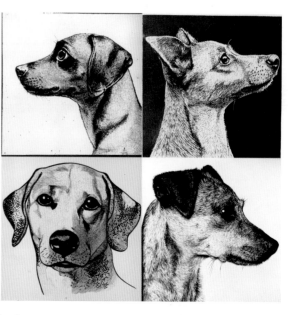

提示詞 ⇒ a dog –sref random（系統隨機生成數值 3294871718）

我們現在另行生成一個全新主題：女人與她的貓，風格參照數值以第一個 3574895463。

風格參照這個功能非常適合有需要系統性的創風格，例如漫畫、繪本等，可以避免風格不一致的狀況。但實際應用還是必須依創作目的進行調整。

提示詞 ⇒ a woman with her cat –sref 3574895463

範例 07 | --stylize \<number\> 或 --s \<number\>

提示詞 ⇒ child's drawing of a puppyt --s \<number\>

--stylize 0

--stylize 50

--stylize 100(預設)

--stylize 250

--stylize 500

--stylize 750

--stylize 1000

我們可以清楚地比較出，從 stylize100 之後所生成的圖像被賦予更強烈的風格，但已經與提示詞中「child's drawing」（孩童的繪畫）產生比較大的偏離。換言之，**提示詞自我的掌握的程度越高，不需要特意提高風格化。**

範例 08 ｜ -tile 連續圖案

提示詞與一般無異，但只要加上參數 --tile，便可以生成四方連續圖案，這個參數對於需要製作貼紙或紡織布料設計特別好用。

提示詞 ⇒ A delicate pattern featuring stylized butterflies in soft pastel colors like lavender, sky blue, and peach, with bright accents on the wings. The design is created in a soft, hand-drawn style. **--tile** --style raw –
精緻的圖案以薰衣草色、天藍色和桃色等柔和色彩的風格化蝴蝶為特色，翅膀上有明亮的點綴。設計採用柔和的手繪風格。

五、重點與建議

學習撰寫提示詞、掌握 Midjourney 的命令及參數，能夠大大提升你在 AI 影像生成平台上的創作效率和作品質量。透過正確且精準的提示詞，AI 可以更快理解你的需求，生成出符合預期的高質量影像。以下是一些重點與建議：

1. 理解 Token 的作用

Midjourney 機器人會將提示詞分解為較小的 token，並將它們與訓練資料對比來生成影像。因此，提示詞越精確、具體，效果就會越好。

2. 簡化提示詞

使用簡單、清晰的短語來描述你想要的內容。避免冗長的句子和複雜的說明。越簡單的描述，AI 越能快速、準確地生成影像。

3. 善用命令與參數

熟悉 Midjourney 的各種命令和參數，比如用「--ar」來調整圖像的縱橫比，或「--cref」來進行人物參照、「--sref」來進行風格參照。這些命令和參數能夠進一步讓生成的影像更符合你的需求。

4. 反覆試驗與調整

提示詞和命令需要不斷地試驗和調整。每次生成影像後，觀察結果並反思如何改進提示詞和命令設置，讓效果更接近你的期望。

5. 利用範例學習

參考書中的範例提示詞，或從 Midjourney 官網和社群中的範例學習，了解其他創作者如何撰寫提示詞和使用命令，這會提供你靈感和實用技巧。

6. 保持創意和靈活性

撰寫提示詞和使用命令時，保持創意和靈活性。AI 影像生成技術提供了豐富的創作可能性，不斷嘗試新想法甚至新物件，突破傳統界限，創作出獨特且令人驚豔的作品。

總結來說，學習和掌握提示詞的撰寫技巧，以及 Midjourney 的命令和參數，能讓你在 AI 影像生成平台上更加得心應手。透過不斷的實踐和創新，你將能夠創作出更加精美和符合期待的影像作品。

午后

PART 3

如何寫好
提示詞

▶ 撰寫提示詞三要訣

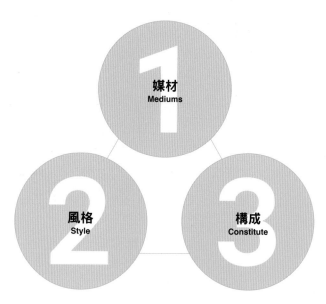

本單元將帶你了解 AI 影像生成的各種元素（條件），如何撰寫提示詞來實現最有效率的創作方式。

1. 媒材

媒材決定了生成影像的質感和視覺效果，是提示詞中最重要的元素之一。

● **範例**：水彩、油畫、數位插畫、鉛筆素描、雕塑。

● **選擇提示**：根據需求描述媒材，例如「Acrylic painting」表達壓克力畫的特性，或「charcoal sketch」指碳筆速寫。

2. 風格

風格為影像添加情感和藝術表現力，能夠定義影像的整體氛圍。

● **藝術風格**：如印象派、超現實主義、立體派。

● **時代風格**：如維多利亞時期、復古 80 年代。

● **流派風格**：龐克、未來主義、極簡主義。

3. 構成

● **構成**是畫面的骨架，描述影像中包含的場景與元素。

● **場景元素**：指定場所（森林、宇宙、城市）或主體（角色、建築、自然景觀）。

● **視角與構圖**：如鳥瞰視角、聚焦主體、動態構圖。

● **燈光與色彩**：添加光影效果（背光、柔光）和色彩氣氛（冷色調、暖色調）。

● **情感與動態**：加入動態（風吹、奔跑）或氛圍（神秘、浪漫）。

運用「媒材、風格、構成」三大要訣，提示詞能更準確表達需求，生成更具美感與一致性的影像。

一、媒材（Mediums）

在 AI 生成圖像的過程中，媒材指的是創作影像時所模擬的藝術風格或工具，也就是影像將以什麼表現形態呈現。例如油畫、水彩、壁畫、雕塑、版畫等等。接下來將使用 ChatGPT（DALL.E）和 Midjourney 生成不同的媒材範例介紹。這些範例的媒材名稱，可以做為日後的參考基礎。

首先在 ChatGPT（DALL.E）上以同樣主題套用不同的媒材，一般習慣上會將媒材名稱，寫在提示詞的最前面，後面才接續主題跟風格及構成要素。

範例

AI 平台 ⇒ ChatGPT（DALL.E）

範例構成｜一隻狗，套用 100 種不同的媒材。 🖥 065

提示詞 ⇒ （介詞名稱），一隻狗（A dog）

01. Woodcut print
木刻版畫

02. Pen sketch
鋼筆素描

03. Watercolor
水彩畫

04. Oil painting
油畫

05. Pencil sketch
鉛筆素描

06. Digital illustration
數位插畫

07. drawing Pastel
粉彩畫

08. Collage
拼貼畫

09. Mosaic art
馬賽克藝術

10. Charcoal drawing
碳筆畫

11. Acrylic painting
壓克力畫

12. Printmaking
版畫

13. Oil pastel
油粉畫

14. Mural
壁畫

15. Colored pencil drawing
彩色鉛筆畫

16. Crayon drawing
蠟筆畫

17. Sketch
素描

18. Enamel painting
琺瑯畫

19. Acrylic paint
丙烯顏料畫

20. Sculpture
雕塑

21. Ceramic art
陶瓷藝術

22. Wire sculpture
鐵線雕塑

23. Fabric collage
布料拼貼

24. Sand painting
砂岩畫

25. Kinetic sculpture
動力雕塑

26. Lacquer painting
漆畫

27. Spray paint art
噴漆藝術

28. Cardboard art
瓦楞紙板藝術

29. Glass painting
玻璃畫

30. Gold leaf painting
金箔畫

31. Stained glass
鉛玻璃畫

32. Stone carving
石雕

33. Bronze sculpture
青銅雕塑

34. Chalk drawing
粉筆畫

35. Lithography
石版畫

36. Plastic art
塑膠藝術

37. Glass sculpture
玻璃雕塑

38. Wood carving
木雕

39. Tapestry
織錦畫

40. Weaving art
編織藝術

41. Calligraphy
書法

42. Block printing
雕版印刷

43. Leathercraft
皮革工藝

44. Embroidery
繡花藝術

45. Digital pixel art
數位像素藝術

46. Silk painting
絲綢繪畫

47. Aluminum foil art
鋁箔藝術

48. Fluorescent painting
螢光粉畫

49. Acid etching
酸蝕版畫

50. Encaustic painting
熔蠟畫

51. Ceramic tile mosaic
瓷磚馬賽克

52. Shadow box art
立體盒藝術

53. Embroidered patch
電繡畫

54. Metalwork
金屬工藝

55. Thread painting
刺繡畫

56. Needlepoint
針織畫

57. Fabric applique
布貼畫

58. Resin art
樹脂藝術

59. Felt art
毛氈藝術

60. Paper quilling art
捲紙藝術

61. Gouache painting
廣告畫

62. Cable sculpture
電纜線雕塑

63. Glass painting
玻璃畫藝術

64. Textile art
紡織藝術

65. Fabric art
布藝藝術

66. Sand art
砂畫

67. Pastel drawing
粉彩畫

68. Crayon drawing
蠟筆畫

69. Sketch
素描

70. Stained glass
鉛玻璃畫

71. Ink wash painting
水墨畫

72. Paper cutting
剪紙藝術

73. Stained wood art
染木藝術

74. Mural art
壁畫藝術

75. Acrylic painting
壓克力畫

76. Digital art
數位藝術

77. Glass mosaic
玻璃馬賽克

78. Textile art
紡織藝術

79. Wire art
線條藝術

80. Chalk art
粉筆畫

81. Encaustic art
熔蠟畫

82. Glass sculpture
玻璃雕塑

83. Lithography
石版畫

84. Snow carving
雪花雕刻

85. Coffee painting
咖啡渲染畫

86. Paper pulp painting
紙漿畫

87. Berry dye painting
漿果染色畫

88. Bread sculpture
麵包雕刻

89. Candy collage
糖果拼貼

90. Plastic weaving
塑膠編織藝術

91. Jade carving
玉石雕刻

92. Rainbow powder painting
噴漆藝術

93. Spin art
旋轉畫

94. Enamel art
瓷漆藝術

95. Bird feather art
鳥羽藝術

96. Bone carving
骨雕藝術

97. Frosted glass painting
磨砂玻璃畫

98. Mold painting
模具畫

99. Circuit board art
電路板藝術

100. Neon carving
霓虹燈雕刻

同樣的，我們也使用 Midjourney 以同主題不同媒材生成不同的影像，讓讀者感受一下不同平台間的差異。對比以上，讀者不難發現 Midjourney 對於藝術類媒材的表現更為精準，更具現實感。

範例

AI 平台 ⇒ Midjourney

範例構成 | 一位戴著黑色眼鏡的光頭男性叼著雪茄為主題，套用數種不同的媒材。　📺 **072**

提示詞 ⇒ （媒材名稱），一位戴著黑色眼鏡的光頭男性叼著雪茄。

An image of Watercolor, a bald man wearing black glasses having a cigar in the style of Watercolor --ar 2:3 --style raw

01. Watercolor
水彩畫

02. Woodcut print
木刻版畫

03. Pen sketch
沾水筆

04. Oil painting
油畫

05. Pencil sketch
鉛筆素描

06. Pastel drawing
粉彩畫

07. Collage
拼貼畫

08. Mosaic art
馬賽克藝術

09. Charcoal drawing
碳筆畫

10. Bone carving
骨雕

11. Circuit board art
電路板藝術

12. Neon carving
霓虹光雕

13. Relief painting
浮雕畫

14. Acrylic painting
壓克力畫

15. Crayon drawing
蠟筆畫

16. Gold leaf painting
金箔畫

17. Metal art
金工藝術

18. 3D puzzle
立體拼圖

19. Computer animation
電腦繪畫

20. 3D origami
折紙藝術

21. Clay art
黏土藝術

二、風格（Style）

在影像生成平台上，「風格」通常指的是一種特定的視覺表現方式，用來塑造和識別圖像的獨特外觀。風格可以受到多種因素的影響，例如：

● **色彩運用**：如明亮的色彩、單色調、對比強烈的色彩等。

● **構圖方式**：如對稱構圖、中心構圖、斜線構圖等。

● **主題選擇**：如自然風光、人像、抽象藝術等。

● **技術手法**：如鏡頭長時間曝光、微距攝影、HDR 攝影等。

● **文化和歷史背景**：如文藝復興風格、巴洛克風格、日本浮世繪等。

風格可以是一種型態、一種視覺效果或為某位藝術家或其特定作品所代表，在影像生成平台上，每個藝術家或作品風格可以被視為一個「Token」。Token 是用來指代特定風格的關鍵字或標籤，AI 模型通 Token 識別並應用相應的風格特徵 Token 可以包含多種元素，例如：

● **色彩使用**：每個藝術家對色彩的運用都有獨特的偏好和技巧。

● **筆觸與紋理**：不同藝術家的筆觸和畫面質感各異。

● **構圖與佈局**：藝術家在畫面中的元素安排和視覺重點有所不同。

● **主題與題材**：藝術家經常關注特定的主題和題材，這也影響其作品的風格。

我們也可以這樣認知，AI 在學習建立影像模型時，給予這些「名詞」相對應的定義，例如：

● **Vincent van Gogh**：強烈筆觸、鮮明色彩。

● **Pablo Picasso**：立體派、幾何構圖。

● **Claude Monet**：印象派、柔和光影。

● **Salvador Dalí**：超現實主義、夢幻場景。

● **Andy Warhol**：波普藝術、大膽色彩。

因此，當你對風格 Token 理解程度越高，就更容易生成你想要的視覺效果。但需注意，這些風格所引用的創作者名稱，**並非代表作者本人或其作品**，而是 AI 平台在模型訓練過程中，分析作品特點後為之名，再取以上的概念再生成成新的影像。

Token 的使用方法，可以根據創作的需求，選擇合適的藝術家風格。例如，如果想要生成具有夢幻色彩的圖像，可以選擇薩爾瓦多‧達利（Salvador Dalí）的風格 Token。

> **提示詞：XXXX 主題，in the style of**（風格名稱）
> 「 **Fantasy color theme ,in the style of Salvador Dalí.** 」

▶ 如何用風格創造作品

使用 AI 模型內建的風格來進行創作，可以說是目前最快速精準的生成方式，我們可以先理解 AI 圖像模型的訓練邏輯。

以梵谷為例：首先，需要收集大量的梵谷作品的影像數據，再透過卷積神經網路（CNN）及生成對抗網路（GAN），提取圖像特徵和分類。CNN 可以學習圖像中的低級特徵（如邊緣、角點）和高級特徵（如形狀、風格）。再透過 GAN 由生成器和判別器組成，生成器創造圖像，判別器判斷圖像的真實性。反覆訓練使生成器能夠產生更加逼真的圖像。

這些模型訓練學習梵谷作品中的色彩運用、筆觸特點和構圖方式。例如其經常使用強烈的色彩對比和旋渦狀的筆觸。那在提示詞的描述中，我們當然可以用最細節的描述，從構圖、筆觸、色彩逐一說明，但現實很難，因為風格更多代表的是一種視覺印象，即便是藝評家都很難鉅細靡遺地用文字描寫出來，但我們現在只需要知道「梵谷」所代表的風格即可。而所謂的風格可以為幾部分：從大範圍到小範圍，我們用個例子來解釋。

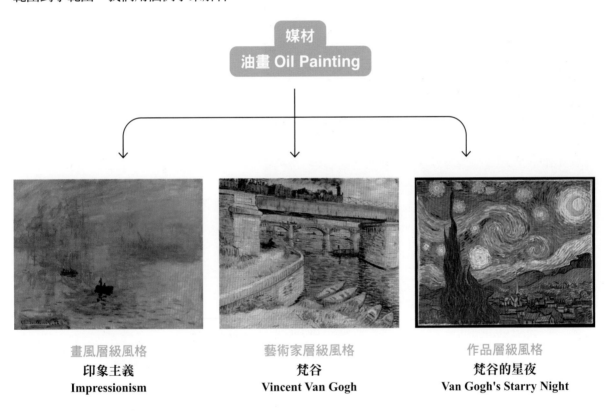

媒材
油畫 Oil Painting

畫風層級風格
印象主義
Impressionism

藝術家層級風格
梵谷
Vincent Van Gogh

作品層級風格
梵谷的星夜
Van Gogh's Starry Night

如果你已經很清楚每位元藝術家甚至作品的特定風格，那麼可以一次就設定風格到作品層級，接下來我們以實際的例子在 Midjourney 上做示範：

> **提示詞：（媒材名稱）+ 主題 + in the style of（風格名稱）**

提示詞 ⇒ an **oil painting** depicting of downtown of Taipei City in the style of **Van Gogh** --style raw --ar 2:3 --v 6.0
油畫，臺北街頭，梵谷風格

提示詞 ⇒ an **oil painting** depicting of downtown of Taipei City in the style of **impressionism** --style raw --ar 2:3 --v 6.0
油畫，臺北街頭，印象主義風格

提示詞 ⇒ an **oil painting** depicting of downtown of Taipei City in the style of **Van Gogh's Starry Night** --style raw --ar 2:3 --v 6.0
油畫，臺北街頭，梵谷星夜風格

▶ 風格與媒材可交叉生成

根據目前各方資料顯示，目前 Midjourney 所建立的風格模型，基本已經含括了西洋美術史和東洋美術史各時期的藝術風格及名家風格，部分像是梵谷可以到作品層級，以 Midjourney 來說，風格參照是很重要的一部分，對於各類藝術及作品風格更清晰，就可省略許多提示詞的描述，只需要具體指出藝術家或作品名稱，AI 便可進行對應的影像生成。

我們將以 Midjourney 陸續生成各種不同媒材、時期及名家的風格，作為讀者後續生成圖像的參考。也請各位讀者切記，「風格化」是 Midjourney 這個平台最具特色的生成邏輯，而風格是可以與媒材交叉應用的。

> 形成一種不同組合的生成趣味。

> 提示詞：（媒材名稱）**+ 主題 + in the style of** （風格名稱）

提示詞 ⇒ a watercolor depicting of downtown of Taipei City in the style of Van Gogh's Starry Night --style raw --ar 2:3 --v 6.0

水彩，臺北街頭，梵谷星夜風格

▶ 介紹八種風格定義

接下來，我們分別提供了繪畫藝術風格、攝影風格、建築風格、插畫漫畫風格、卡通動畫風格、電影風格等數百位代表人物的藝術定義，並透過 Midjourney 進行生成，提供給大家做為創作前參考學習。

善用風格定義，可以大幅縮短對於影像美學的學習，但長期而言，我們仍是鼓勵所有人對於視覺的構成，包括構圖、色彩、主題元素、表現技法等都能理出一個屬於自己的邏輯，但站在巨人的肩膀上學習，就如同學畫者必從臨摹開始是同一個道理。

01 繪畫藝術風格 🖥 078

使用 Midjourney 或其他平台生成影像時，理解不同繪畫風格的多樣性和特徵是至關重要的。繪畫風格不僅表現了藝術家的情感和意圖，也決定了影像如何傳達資訊和氛圍。每種風格都有其獨特的視覺語言和美學價值，這對於 Midjourney 這類 AI 工具在生成符合預期的影像非常重要。

例如，選擇巴洛克風格時，理解其華麗、動感和豐富裝飾的特點可以引導 AI 生成出充滿活力和戲劇性的影像。同樣，對於現代主義風格，掌握其簡約、功能主義和強調幾何形狀的特性，可以幫助生成出清晰、乾淨且富有現代感的設計。

此外，理解繪畫風格也有助於確保生成的影像在風格一致。例如，當一個項目需要統一的視覺風格時，正確運用同一種繪畫風格可以避免風格混亂，從而提升作品的整體品質和專業性。

所以，深入理解繪畫風格是使用 Midjourney 等 AI 工具創作的基礎。這不僅有助於更好地控制和引導 AI 的創作過程，也能更準確地實現創作者的藝術意圖和視覺目標。在本單元中，我們將介紹不同藝術發展時期的代表藝術家及其作品，並透過 AI 生成的例子，方便日後進行索引和參閱。

中世紀時期（約 5 世紀 -15 世紀）

引用中世紀（約 5 世紀 -15 世紀）六位代表藝術家的風格，以相同提示詞生成影像。

相同提示詞 ⇒ 一個小女孩與她的貓。

使用說明 ⇒ 在同一段提示詞中，可透過參數**雙花括弧** {{ 變因 1, 變因 2, 變因 3,⋯⋯變因 10}} 設定，加入不同的變因，最多可以添加至十組變因，以「,」隔開，可以一次性將所有變因獨立生成。

An artwork depicting of a little girl with a cat in the style of {{ Hildegard of Bingen, Ambrogio Lorenzetti, Giotto di Bondone, Duccio di Buoninsegna, Jan van Eyck, Andrei Rublev. }} --style raw

01. Hildegard of Bingen
希爾德加德・馮・賓根

神秘的宗教插畫，展現靈性的啟示與精神探索。

02. Ambrogio Lorenzetti
安布羅焦・洛倫澤蒂

用和諧的色彩與細緻的透視，描繪社會理想與人性光輝。

03. Giotto di Bondone
喬托・迪・邦多納

獨特的宗教敘事與人物表現，開創了文藝復興繪畫的空間深度。

04. Duccio di Buoninsegna
杜喬・迪・布奎尼亞

柔和的色彩與典雅的構圖，展現拜占庭風格的神聖之美。

05. Jan van Eyck
揚・范・艾克

以精緻的細節與光影運用，塑造出真實的北方文藝復興風格。

06. Andrei Rublev
安德列・魯布列夫

金色背景與莊嚴構圖，展現東正教聖像畫的精神深度。

文藝復興時期

文藝復興時期，列舉該時期具有代表性的藝術家，以同樣的提示詞，生成不同的風格的作品。

相同提示詞 ⇒ 女性肖像畫。

a female portrait painting in the style of {{Leonardo da Vinci, Raphael Sanzio,Michelangelo Buonarroti, Titian, Sandro Botticelli, Jacopo de' Barbari, Federico Zuccari, Pietro Perugino, Andrea Mantegna, Lorenzo Lotto}} -ar 2:3 --style raw --v 6.0

透過這樣的練習，可以比較容易理解各個藝術家透過 AI 表現後的風格差異，爾後在生成影像時，也可根據不同的主題搭配不同的藝術家風格。

但有一點必須要注意，並不是所有的主題任意套用藝術家風格都可以得到滿意的效果，必須要認知

01. Leonardo da Vinci
李奧納多・達文西

將藝術與科學結合，展現
人類解剖與神秘氛圍的奇
妙融合。

02. Raphael Sanzio
拉斐爾・聖齊奧

透視精確、人物和諧，塑
造了理想主義與人文精神
的完美結合。

03. Titian
提香・韋切利奧

濃烈的色彩與戲劇性的構
圖，推動了威尼斯畫派的
藝術風格。

04. Sandro Botticelli
桑德羅・波提切利

浪漫與神話的融合，輕盈
的線條展現夢幻之美。

05. Federico Zuccari
費德里科・費利尼

強烈的動態與戲劇性構
圖，塑造晚期風格主義的
藝術語彙。

06. Pietro Perugino
彼得羅・佩魯吉諾

和諧色調與流暢線條，呈
現靜謐與神聖的氛圍。

07. Andrea Mantegna
安德列亞・曼特尼亞

精確透視與歷史敘事，賦
予宗教史詩般的莊嚴。

08. Lorenzo Lotto
洛倫佐・洛托

內省的情感與個性化的肖
像，展現人性深處的矛盾。

09. Michelangelo Buonarroti
米開朗基羅・博那羅蒂

力量與張力的雕塑與繪畫，展現人類精
神的崇高之美。

10. Jacopo de' Barbari
雅克布・德・巴巴里

精確的透視與數學結構，展現文藝復興
的理性之美。

所謂藝術家風格也是基於以往的作品進行模型訓練，如達文西，過去的作品主要是人物，我們用其風格生成各種人物，可以得到不錯的效果，但若用達文西風格生成太空戰艦（如下圖）是比較難判斷是否貼近期待中的風格，有時不妨多嘗試這種跨度大的影像生成，也許會得到意想不到的效果。

此外，在套用藝術家風格時，若提示詞並未充分形容時空背景，則 AI 會自動依藝術家當時的時空背景進行生成，如前面所生成中世紀藝術家的影像，AI 會自動模擬當時繪製在牆體上，存至今日牆體剝落的視覺效果。或是文藝復興時期多數繪畫在木板上，現今油彩剝落產生的紋路或裂痕。

達文西風格的太空船
提示詞 ⇒ a space alien ship painting **in the style of Leonardo da Vinci** --ar 3:2 --style raw --v 6.0

巴洛克時期（約 17 世紀 -18 世紀）

文藝復興之後緊接著進入到巴洛克時期（約 17 世紀 -18 世紀），範例將以貓王作為創作主題，並以十位代表性藝術家進行影像生成。
這單元的練習，將根據不同藝術家風格在提示詞中加入更多時空背景元素，營造出各種不同風貌的貓王印象。

01. Peter Paul Rubens
彼得・保羅・魯本斯

奢華的構圖與戲劇性的光影，塑造出巴洛克的視覺盛宴。

A dynamic portrait in the style of Peter Paul Rubens, depicting Elvis Presley. The portrait captures Elvis in a lively, expressive pose, with vibrant colors and dramatic lighting. The background features a vibrant, mythological landscape with dynamic skies, typical of Rubens' robust and energetic Baroque style. The light is golden and dramatic, highlighting the robust forms and vibrant colors. --ar 2:3 --style raw --v 6.0

02. Caravaggio
卡拉瓦喬

強烈的光影對比與真實主義，捕捉生命的戲劇瞬間。

A dramatic portrait in the style of Caravaggio, showing Elvis Presley. The portrait is set in a dimly lit, intimate setting, with Elvis's face illuminated by a single, intense light source, creating stark contrasts and deep shadows. The scene captures the realistic, almost theatrical drama typical of Caravaggio's late 16th to early 17th-century works. --ar 2:3 --style raw --v 6.0

03. Diego Velázquez
維拉斯奎茲

精確的光影與貴族肖像，揭示權力與人性的深層矛盾。

Arealistic portrait in the style of Diego Velázquez, depicting Elvis Presley in a royal court setting. Elvis is shown in a natural, relaxed pose, with soft, diffused lighting and a muted color palette, capturing the refined elegance and detailed realism typical of Velázquez's 17th-century works. The background features a subtle, regal ambiance. --ar 2:3 --style raw --v 6.0

04. Rembrandt van Rijn
倫勃朗・凡・萊茵

深邃的光影與細膩的情感，描繪出靈魂深處的戲劇性。

A dramatic portrait in the style of Rembrandt van Rijn, showing Elvis Presley in a dimly lit room with rich, warm tones. Elvis's expressive face is illuminated by Rembrandt's signature use of chiaroscuro, creating deep contrasts and an emotional depth. The background features dark, richly textured walls, adding to the intimate and contemplative atmosphere. --ar 2:3 --style raw --v 6.0

05. Gian Lorenzo Bernini
詹尼・洛倫佐・貝爾尼尼

充滿動態與情感的雕塑，賦予大理石生命的律動。

A dynamic sculpture-inspired portrait in the style of Gian Lorenzo Bernini, depicting a lifelike and expressive portrait of Elvis Presley. The scene is set in a grand, Baroque hall with dramatic lighting highlighting the intricate details and lifelike expression, capturing the theatricality and emotion typical of Bernini's mid-17th-century works. --ar 2:3 --style raw --v 6.0

06. Anthony van Dyck
安東尼・范戴克

高貴的肖像與優雅的筆觸，捕捉貴族的威儀與魅力。

A refined portrait in the style of Anthony van Dyck, showing Elvis Presley in an elegant, aristocratic setting. Elvis is portrayed with a sense of nobility and poise, dressed in luxurious, period-appropriate attire. The scene is characterized by soft, natural lighting and a rich color palette, capturing the grandeur and elegance typical of van Dyck's portraits. --ar 2:3 --style raw --v 6.0

07. Francisco de Zurbarán
法蘭西斯科・德・祖巴蘭

簡潔莊重的宗教畫，營造出神秘與靜謐的氛圍。

A dramatic portrait in the style of Francisco de Zurbarán, depicting Elvis Presley in a solemn, monastic setting. Elvis is illuminated by a single, strong light source, creating deep shadows and a sense of spiritual intensity. The background features austere, yet richly textured walls, typical of Zurbarán's early 17th-century works. --ar 2:3 --style raw --v 6.0

08. Francesco Borromini
法蘭西斯科・波羅米尼

複雜的建築設計與大膽的線條，挑戰古典建築規範。

A grand portrait in the style of Francesco Borromini, showing Elvis Presley in a stunning, Baroque architectural setting. Elvis is framed by dramatic, curvilinear forms and intricate details of Borromini's architecture. The light is bright and clear, enhancing the architectural splendor and the refined elegance of the portrait. --ar 2:3 --style raw --v 6.0

09. Claude Lorrain
克勞德・洛蘭

柔和的光線與田園景色，創造出詩意的自然美。

A serene portrait in the style of Claude Lorrain, depicting Elvis Presley in a pastoral, classical landscape. Elvis is surrounded by lush greenery and ancient ruins, with a softly glowing sunset in the background. The light is golden and diffused, capturing the tranquil beauty and timelessness typical of Lorrain's 17th-century works. --ar 2:3 --style raw --v 6.0

10. Giovanni Battista Tiepolo
喬瓦尼‧巴提斯塔‧提也波洛

壯麗的壁畫與戲劇性場景，營造出華麗的視覺奇觀。

A vibrant portrait in the style of Giovanni Battista Tiepolo, showing Elvis Presley in a grand, opulent hall. Elvis is depicted with dynamic poses and expressive features, dressed in elaborate, colorful garments. The scene is filled with bright, luminous colors and dramatic lighting, capturing the theatricality and grandeur typical of Tiepolo's 18th-century works. --ar 2:3 --style raw --v 6.0

我們也可利用上述的提示詞，自行替換主題人物，例如在上面第十個提示詞，將貓王替換成泰勒絲。此外，目前所有的名人，或者說只要維基百科有登錄的，過去有龐大影像資料的名人肖像，基本都已被 AI 平台定義為一個個符元（Token），也就是說只要輸入名字，就可以得到相對應類似的影像。這在某個程度上對於一些商業創作是很方便的。但某些平台也礙於一些政策，將一些名人其姓名列為禁語，避免生成出一些誤導性的影像。而美中不足的是，目前各大模型對於華語世界的史料是很缺乏的，例如 AI 可以識別泰勒絲，但目前卻無法直接對「鄧麗君」進行描繪。類似的狀況還包括藝術家風格、地名、文學作品等等，所以如果要創作中文地區的相關題材，則必須要對內容主題作出鉅細靡遺的描述。這在後面章節範例時會在補充更多的作法。

巴洛克時期結束，出現了新古典主義與浪漫主義（約 18 世紀末 -19 世紀中期），在這階段出現了許多以歐洲帝國及王朝為背景的繪畫作品，以下我們仍列舉數位具代表性的藝術家，並使用她的創作風格進行生成，主題為「騎在白馬上的公主」。

A vibrant portrait in the style of Giovanni Battista Tiepolo, showing **Taylor Swift** in a grand, opulent hall. **She** is depicted with dynamic poses and expressive features, dressed in elaborate, colorful garments. The scene is filled with bright, luminous colors and dramatic lighting, capturing the theatricality and grandeur typical of Tiepolo's 18th-century works. --ar 2:3 --style raw --v 6.0

新古典主義與浪漫主義

本書所有的提示詞，中文部分可以直接應用在像是 ChatGPT, DALL.E 或 Copilot 上，下圖為中文提示詞，以 DALL.E 生成的效果。

01. Jacques-Louis David │ 賈克 - 路易‧大衛

體現共和理想，強調英雄氣概與歷史敘事。

A detailed painting in the style of Jacques-Louis David, depicting a princess in a flowing, neoclassical gown riding a majestic white horse. The scene is set in an opulent palace courtyard with grand, classical architecture in the background. The light is dramatic and crisp, highlighting the princess's regal demeanor and the horse's elegant stance, capturing the grandeur and formality typical of David's late 18th to early 19th-century works. --style raw --ar 1:1 --v 6.0

一幅以賈克 - 路易‧大衛風格描繪的詳細畫作，展示了一位身穿飄逸新古典風格長袍的公主騎著一匹雄偉的白馬。場景設置在奢華的宮殿庭院中，背景是宏偉的古典建築。戲劇性而清晰的光線突出了公主的高貴氣質和白馬的優雅姿態，捕捉了大衛 18 世紀末到 19 世紀初作品中典型的巨集偉和正式感。

02. Jean-Auguste-Dominique Ingres │ 尚‧奧古斯特‧多明尼克‧安格爾

細膩的肖像與優雅的線條，展現古典主義之美。

A detailed painting in the style of Jean-Auguste-Dominique Ingres, depicting a princess in an intricately detailed, elegant gown, seated gracefully on a white horse. The scene is set in a lush, verdant garden with classical sculptures and fountains. The light is soft and refined, emphasizing the smooth textures and delicate features, characteristic of Ingres' early 19th-century works. --style raw --ar 1:1 --v 6.0

我們再同樣以中文提示詞以 DALL.E 生成，大家可以比較一下二者間的差異。由於我個人更偏好 Midjourney 的生成效果，因此以下範例就都由 Midjourney 生成，但中文提示詞是可以適用於其他平台的。

03. Eugène Delacroix
德拉克羅瓦

鮮明色彩與動感構圖，表現戲劇性的浪漫主義風格。
A vibrant painting in the style of Eugène Delacroix, showing a princess in a richly colored, dramatic gown riding a powerful white horse. The scene is set in a dynamic, natural landscape with dramatic skies and lush foliage. The light is intense and contrasting, highlighting the movement and emotion, reflecting Delacroix's romantic and energetic style from the mid-19th century. --ar 1:1 --style raw --v 6.0

04. Francisco Goya
法蘭西斯科・哥雅

深刻的視覺表達，揭示社會與人性的黑暗面。
A dramatic painting in the style of Francisco Goya, depicting a princess in a dark, ornate gown, riding a white horse in a moonlit, eerie forest. The scene is filled with shadows and mysterious elements, capturing the haunting and introspective mood typical of Goya's late 18th to early 19th-century works. --ar 1:1 --style raw --v 6.0

05. J.M.W. Turner
威廉・透納

光影變化與自然力量，創造出戲劇性的浪漫景觀。
A luminous painting in the style of J.M.W. Turner, showing a princess in a flowing, ethereal gown riding a white horse through a misty, atmospheric landscape. The scene features a dramatic sunset with golden light and swirling clouds, capturing the sublime beauty and luminous quality typical of Turner's early 19th-century works. --ar 1:1 --style raw --v 6.0

06. John Constable
約翰‧康斯特勃

捕捉田園風光，展現自然的寧靜與詩意。

A picturesque painting in the style of John Constable, depicting a princess in a simple, yet elegant gown riding a white horse through a tranquil, pastoral landscape. The scene features rolling hills, rustic cottages, and lush greenery under a bright, clear sky, with soft, natural light emphasizing the serene beauty of the English countryside, characteristic of Constable's early 19th-century works. --ar 1:1 --style raw --v 6.0

07. Caspar David Friedrich
卡斯帕‧大衛‧弗裡德里希

荒野與孤獨的場景，傳達靈性冥想與超越。

A hauntingly beautiful painting in the style of Caspar David Friedrich, depicting a princess in a flowing, mystical gown riding a white horse through a foggy, mysterious forest. The scene is set at dawn or dusk, with ethereal light filtering through the trees, capturing the sense of solitude and introspection typical of Friedrich's early 19th-century works. --ar 1:1 --style raw

08. Pierre-Paul Prud'hon
皮埃爾-保羅‧普呂東

柔和的光線與古典美學，傳遞深沉的情感。

A romantic painting in the style of Pierre-Paul Prud'hon, showing a princess in an elegant, flowing gown riding a white horse in a moonlit, enchanted forest. The scene features delicate, soft lighting and gentle shadows, creating a dreamy, intimate atmosphere typical of Prud'hon's late 18th to early 19th-century works. --ar 1:1 --style raw --v 6.0

09. Camille Corot
卡米耶‧科羅

田園景色中的光與影，呈現出自然的靜謐與和諧。

A serene painting in the style of Camille Corot, depicting a princess in a simple, yet elegant gown riding a white horse through a peaceful, verdant landscape. The scene features soft, diffuse light and delicate, natural tones, capturing the tranquil beauty of nature typical of Corot's mid-19th-century works. --ar 1:1 --style raw --v 6.0

10. Henri Fuseli
約翰‧亨希‧菲斯利

奇幻與戲劇性的場景，探索潛意識的夢境世界。

A dramatic painting in the style of Henri Fuseli, depicting a princess in a dark, flowing gown riding a white horse through a surreal, nightmarish landscape. The scene features dramatic, contrasting light and shadow, with fantastical, eerie elements and intense emotions, reflecting Fuseli's late 18th to early 19th-century works. --ar 1:1 --style raw --v 6.0

以上這十則示例在提示詞的撰寫上，雖然主題都是公主騎白馬，但為了使讀者更暸解原創藝術家的風格，我們在提示詞中賦予了不同的情境，包括公主的表情、服飾及背景。

同樣地，也可嘗試在同個提示詞中，將關鍵字的公主、白馬換成別的主題，例如：男孩、黑熊。我們以約翰‧亨利希‧菲斯利的風格為例，現實中男孩當然不會騎黑熊，但透過 AI，我們得以將藝術家風格進行無限的延伸，進而實現我們的創意。

A dramatic painting in the style of Henri Fuseli, depicting a little boy in a dark, flowing gown riding a black bear through a surreal, nightmarish landscape. The scene features dramatic, contrasting light and shadow, with fantastical, eerie elements and intense emotions, reflecting Fuseli's late 18th to early 19th-century works. --ar 1:1 --style raw --v 6.0

現實主義與印象主義（約 19 世紀中期 -19 世紀末）

在這個階段來到大家比較熟知的一些藝術家風格，如果對於美術史沒有研究，通稱「油畫」，這也是我們在很多平台上看到一些使用者詠唱「咒語：油畫」卻老是得不到理想的效果，要知道油畫藝術發展迄今，先輩藝術家們經過不斷地淬鍊改進，至少已經發展出數百種成熟的風格技巧，又豈是「油畫」二字能夠輕易含括，在以下的單元，我們仍舊列舉出一些代表性的藝術風格，以同樣主題進行生成，讓者便於比較。

1

01. Gustave Courbet
古斯塔夫・庫爾貝

寫實主義的開創者，真實描繪社會與農村生活。

A painting in the style of Gustave Courbet, depicting a father and son in a realistic, rural setting. They are standing in a field with lush greenery, the father wearing work clothes and the son in simple, rustic attire. The background features rolling hills and a distant farmhouse, with natural light casting strong, clear shadows. The scene captures the everyday life and honest, unidealized portrayal of rural people, typical of Courbet's mid-19th century works. --ar 2:3 --style raw --v 6.0

02. Édouard Manet
愛德華・馬奈

打破傳統繪畫法則，探索現代主題與視覺創新。

A painting in the style of Édouard Manet, depicting a father and son in an urban Parisian setting during the late 19th century. They are seated at an outdoor café, the father in a modern suit and hat, the son in a sailor suit. The background features bustling streets and elegant buildings, with soft, diffused light highlighting the figures. The scene captures the modernity and casual elegance typical of Manet's works. --ar 2:3 --style raw --v 6.0

2

03. Claude Monet
克勞德 · 莫內

捕捉瞬間光影與色彩，開創印象派藝術的新視野。

A painting in the style of Claude Monet, depicting a father and son in a vibrant, impressionistic garden. They are standing near a pond filled with water lilies, the father in a light summer suit and the son in a white dress shirt and shorts. The background features lush foliage and colorful flowers, with dappled sunlight creating a shimmering effect. The scene captures the beauty and fleeting moments of nature, typical of Monet's late 19th to early 20th century works. --ar 2:3 --style raw --v 6.0

04. Edgar Degas
愛德格 · 德加

聚焦芭蕾舞者與日常生活，展現動態與細節。

A painting in the style of Edgar Degas, depicting a father and son at a ballet rehearsal in an elegant, dimly lit studio. The father, dressed in a formal suit, is observing while the son, in casual attire, mimics the dancers. The background features ballerinas practicing, with soft, directional light creating dramatic shadows. The scene captures the movement and grace of the dancers, alongside the intimate bond between father and son, typical of Degas' late 19th-century works. --ar 2:3 --style raw --v 6.0

05. Pierre-Auguste Renoir
皮埃爾 - 奧古斯特 · 雷諾阿

柔和的光線與快樂的場景，呈現溫馨與愉悅。

A painting in the style of Pierre-Auguste Renoir, depicting a father and son enjoying a sunny afternoon in a lush park. The father is seated on a bench, dressed in light summer clothing, while the son, in a playful outfit, is running around. The background features vibrant greenery, blooming flowers, and softly filtered sunlight. The scene captures the warmth and joy of family life, typical of Renoir's late 19th to early 20th century works. --ar 2:3 --style raw --v 6.0

06. Camille Pissarro
卡米耶・畢沙羅

捕捉鄉村風光與自然的四季變化，展現平靜之美。

A painting in the style of Camille Pissarro, depicting a father and son walking through a bustling market street. The father is dressed in simple, working-class attire, and the son in a short jacket and cap. The background features market stalls, townsfolk, and charming old buildings, with diffused natural light capturing the lively atmosphere. The scene reflects the everyday life and community spirit typical of Pissarro's late 19th to early 20th century works. --ar 2:3 --style raw --v 6.0

07. Alfred Sisley
阿爾弗雷德・西斯萊

擅長河流與田園景觀，光影變幻令人驚嘆。

A painting in the style of Alfred Sisley, depicting a father and son walking along a serene riverbank. The father, dressed in casual, late 19th-century clothing, and the son in a sailor suit, are strolling under a canopy of trees. The background features the gentle flow of the river, reflections in the water, and a clear, bright sky. The scene captures the peacefulness and natural beauty typical of Sisley's works. --ar 2:3 --style raw --v 6.0

08. Ferdinand Hodler
費迪南德・霍德勒

強調構圖的紀律與象徵主義，探索靈性主題。

A painting in the style of Ferdinand Hodler, depicting a father and son in a majestic mountain landscape. The father, dressed in traditional Swiss attire, stands with his son, who is also in rustic clothing. The background features towering peaks, lush meadows, and dramatic lighting. The scene emphasizes the unity and strength of the human figures against the grandeur of nature, typical of Hodler's early 20th-century works. --ar 2:3 --style raw --v 6.0

09. Mary Cassatt
瑪麗・卡薩特

捕捉親密的家庭場景與女性日常，展現溫柔情感。
A painting in the style of Mary Cassatt, depicting a father and son in a cozy, domestic interior. The father, dressed in a casual suit, is seated in a comfortable chair, reading a book, while the son, in a sailor suit, sits on his lap. The background features soft furnishings and warm, diffused light. The scene captures the intimate, nurturing bond between parent and child, typical of Cassatt's late 19th to early 20th century works. --ar 2:3 --style raw --v 6.0

10. Federico Zandomeneghi
費德里科・薩卡

鮮明的色彩與都市風情，刻畫出市民生活的多樣面貌。
A painting in the style of Federico Zandomeneghi, depicting a father and son in a stylish Parisian park. The father is dressed in elegant, late 19th-century attire, and the son in a charming outfit with a cap. The background features vibrant flower beds, manicured lawns, and fashionable park-goers. The light is bright and lively, capturing the elegance and modernity of the scene, typical of Zandomeneghi's works. --ar 2:3 --style raw --v 6.0

在這個單元中，我們可以看到相同「父與子」主題，以不同風格的藝術家表現，在提示詞中，定義不同的場景及人物型態，便可將創作的範圍無限放大。我們也可試著改變關鍵字，看看會得到甚麼樣影像？

「母與女」主題

A painting in the style of Ferdinand Hodler, depicting a **Mother** and a daughter in a majestic mountain landscape. **The Mother**, dressed in traditional Swiss attire, stands with **her daughter**, who is also in rustic clothing. The background features towering peaks, lush meadows, and dramatic lighting. The scene emphasizes the unity and strength of the human figures against the grandeur of nature, typical of Hodler's early 20th-century works. --ar 2:3 --style raw --v 6.0

後印象主義與象徵主義（約 19 世紀末 -20 世紀初）

繼印象主義之後，繪畫藝術的發展更形多元，這時的風格更多變，我們稱之為後印象主義與象徵主義（約 19 世紀末 -20 世紀初），列舉十位代表性藝術家，並以其風格生成主題為「閱讀的女子」。

01. Paul Cézanne
保羅・塞尚

穩重的構圖與幾何形體，奠定現代藝術的基石。

A painting in the style of Paul Cézanne, depicting a woman reading in a serene, sunlit orchard. She sits on a wooden bench under a fruit-laden tree, wearing a simple elegant gown typical of the late 19th century. The dappled sunlight filters through the leaves, casting soft shadows. The background features the rolling hills of Provence, creating a tranquil and timeless atmosphere. --ar 2:3 --style raw --v 6.0

02. Vincent Van Gogh
梵谷

激情的筆觸與強烈的色彩，表現內心的情感與掙扎。

A vibrant painting in the style of Vincent van Gogh, showing a woman reading in a cozy, rustic room filled with swirling light. She is dressed in a simple, flowing dress, seated by a window with sunlight streaming in, illuminating her golden hair and book. The room is painted with van Gogh's bold, swirling brushstrokes and rich, vibrant colors, capturing an intense, emotional atmosphere. --ar 2:3 --style raw --v 6.0

03. Paul Gauguin
保羅‧高更

探索異國風情與精神世界，運用大膽的色彩與簡化的形式。

A colorful painting in the style of Paul Gauguin, depicting a woman reading in a lush, tropical garden. She is dressed in traditional Tahitian attire, sitting on a woven mat surrounded by vibrant flowers and exotic plants. The background features a thatched hut and distant mountains, with warm, golden light of the late afternoon sun casting long shadows. Gauguin's flat, bold colors and simplified forms highlight the scene's exotic beauty. --ar 2:3 --style raw --v 6.0

04. Georges Seurat
喬治‧修拉

精確的點彩技法與色彩理論，創造視覺震撼的作品。

A pointillist painting in the style of Georges Seurat, showing a woman reading in a bustling Parisian park. She is dressed in a fashionable, light-colored dress, sitting on a blanket under a large tree. The background is filled with other park-goers enjoying the sunny day, depicted with Seurat's characteristic dots of pure color. The bright, even light captures the lively, yet tranquil atmosphere of the park. --ar 2:3 --style raw --v 6.0

05. Henri Rousseau
亨利‧盧梭

奇幻與天真的畫風，描繪夢幻般的異想世界。

A whimsical painting in the style of Henri Rousseau, depicting a woman reading in a fantastical jungle setting. She is dressed in an elegant gown, sitting on a tree stump surrounded by dense foliage, exotic animals, and vibrant flowers. The light filters through the canopy, creating a magical, dappled effect. The scene combines realistic detail with a dreamlike quality, reflecting Rousseau's naive, yet richly imaginative style. --ar 2:3 --style raw --v 6.0

06. Edvard Munch
愛德華・蒙克

揭示人類情感的深層痛苦與孤獨，創造心理暗示的視覺語言。

A dramatic painting in the style of Edvard Munch, showing a woman reading in a dimly lit room. She is dressed in dark, flowing clothing, seated at a small table with a single lamp casting a warm, haunting glow. The background features swirling, abstract patterns and muted colors, capturing a sense of melancholy and introspection typical of Munch's early 20th-century works. --ar 2:3 --style raw --v 6.0

07. Odilon Redon
奧迪隆・雷東

神秘與幻想主題，探索潛意識的未知領域。

A mystical painting in the style of Odilon Redon, depicting a woman reading in a surreal, dreamlike environment. She is dressed in ethereal, flowing garments, sitting on a floating rock surrounded by fantastical creatures and delicate flowers. The background is filled with soft, luminous colors and intricate, otherworldly details, creating a sense of wonder and enchantment. --ar 2:3 --style raw --v 6.0

08. Gustav Klimt
古斯塔夫・克林姆特

華麗的裝飾風格與金箔運用，展現夢幻與情慾之美。

An intricate and elegant illustration of a woman reading, drawn in the style of Gustav Klimt. The woman is depicted sitting gracefully, immersed in a book, with her flowing dress adorned with Klimt's signature decorative patterns and gold leaf accents. The background features a richly detailed and ornate design, incorporating floral motifs and abstract shapes, enhancing the luxurious and serene atmosphere. The overall scene captures the essence of tranquility and intellectual engagement, with vibrant colors and delicate textures characteristic of Klimt's artistic style. --ar 2:3 --style raw

09. James Ensor
詹姆斯・恩索爾

怪誕與諷刺的圖像，揭露社會的荒謬與虛偽。

A surreal painting in the style of James Ensor, depicting a woman reading in a chaotic, intriguing setting. She is dressed in early 20th-century attire, sitting at a table surrounded by bizarre masks and strange figures. The background features a cluttered room with vibrant, clashing colors and eerie shadows, reflecting Ensor's fascination with the grotesque and fantastical. The light is dramatic, casting sharp contrasts and adding to the surreal atmosphere. --ar 2:3 --style raw --v 6.0

10. Pierre Bonnard
皮埃爾・波納爾

柔和的色彩與親密的家庭場景，散發溫馨與詩意之美。

A warm, intimate painting in the style of Pierre Bonnard, showing a woman reading in a sunlit, domestic interior. She is dressed in a simple, elegant dress, sitting on a comfortable sofa with sunlight streaming through large windows. The background features a cozy room with soft, muted colors and Bonnard's characteristic attention to light and texture, capturing the quiet, serene beauty of everyday life in the early 20th century. --ar 2:3 --style raw --v 6.0

在這幾段提示詞的撰寫中，除了主題外，我們加入了許多該藝術家的風格特色描寫，或是其擅長的景物表現，這些資料目前都可以在網路上非常容易取得，我們在撰寫提示詞時，可以多參考這些外部資料。

另外一方面，由於 Midjourney 的模型訓練已經是以藝術家風格做於來源，更簡單的做法是只要指定藝術家風格，其餘就讓 AI 自行發揮，這種提示詞的寫法有好有壞，最直接的缺點當然就是少了參與感，畢竟那不是你的想法或是觀察，但有時當缺乏靈感時，也就不妨讓 AI 自行發揮，成為創作的靈感參考。

我們再另做一個風格交錯的實驗，同樣是閱讀的女子為主題，風格設定為高更及克林姆，看看會有甚麼樣的驚喜。

An intricate and elegant painting of a woman reading, in the style of Paul Gauguin. in the style of Gustav Klimt. --ar 2:3 --style raw --v 6.0

這是一幅複雜而優雅的繪畫，描繪了一位正在閱讀的女人，具有保羅·高更的風格。古斯塔夫·克林姆特的風格。

很明顯看出高更的用色以及克林姆的技法，實際上我們無法跟這二位大師巨擘合作創作，但透過 AI，我們發掘了更多創作的可能性。

現代主義（約 20 世紀初 -20 世紀中期）

很快地時間來到現代主義（約 20 世紀初 -20 世紀中期），這個時期有更多我們熟悉的藝術家，像是畢卡索、蒙德里安等等，以下我們將以「歐洲王室貴族與日本藝妓」為題，以現代主義代表性風格生成作品，前段提示詞均相同，但後段則根據不同藝術家風格再進行細部描述。

01. Pablo Picasso
巴勃羅‧畢卡索

立體派創始人，突破傳統，探索多樣藝術風格。
A detailed image in the style of Pablo Picasso, depicting a European medieval prince and a beautiful Japanese geisha. The composition uses Picasso's iconic Cubist elements, with fragmented forms and multiple perspectives. The prince is adorned in medieval attire with geometric shapes and abstract angles, while the geisha is depicted with traditional kimono patterns broken into colorful, angular segments. The background features a blend of European and Japanese architecture, rendered in a cubist manner, emphasizing the fusion of cultures. The overall palette is bold and vibrant, reflecting Picasso's experimental and avant-garde approach. --ar 2:3 --style raw --v 6.0

02. Henri Matisse
亨利‧馬蒂斯

色彩大師，以大膽的色塊與裝飾性構圖著稱。
A detailed image in the style of Henri Matisse, depicting a European medieval prince and a beautiful Japanese geisha. The scene is characterized by Matisse's use of bright, bold colors and simplified, flowing forms. The prince is dressed in medieval attire with rich, expressive colors, and the geisha is adorned in a vibrant kimono with large, decorative patterns. The background features a stylized blend of European and Japanese elements, with fluid lines and a harmonious composition. The overall mood is joyous and lively, capturing Matisse's Fauvist style. --ar 2:3 --style raw --v 6.0

03. Wassily Kandinsky
瓦西裡 · 康丁斯基

抽象藝術先驅,運用色彩與形狀表現音樂性。

A detailed image in the style of Wassily Kandinsky, depicting a European medieval prince and a beautiful Japanese geisha. The composition uses Kandinsky's abstract and geometric forms, with a focus on color harmony and dynamic movement. The prince and geisha are represented with fluid, interlocking shapes and vibrant colors. The background is an abstract blend of European and Japanese architectural elements, with swirling lines and bold contrasts. The overall effect is a vibrant and energetic depiction, reflecting Kandinsky's abstract expressionism. --ar 2:3 --style raw --v 6.0

04. Marcel Duchamp
馬塞爾 · 杜尚

現代藝術的顛覆者,以現成物挑戰傳統藝術界限。

A detailed image in the style of Marcel Duchamp, depicting a European medieval prince and a beautiful Japanese geisha. The scene incorporates Duchamp's Dadaist elements, with a playful and thought-provoking composition. The prince is adorned in medieval attire with a mix of realistic and abstract features, while the geisha is depicted with traditional kimono patterns and surreal elements. The background features a blend of European and Japanese architecture, with unexpected juxtapositions and visual puns. The overall style is innovative and avant-garde, reflecting Duchamp's approach to art. --ar 2:3 --style raw --v 6.0

05. Joan Miró
胡安 · 米羅

夢幻般的超現實主義,運用符號與鮮豔色彩。

A detailed image in the style of Joan Miró, depicting a European medieval prince and a beautiful Japanese geisha. The composition uses Miró's signature whimsical and surreal elements, with playful shapes and vibrant colors. The prince and geisha are rendered with simple, abstract forms, and bold, childlike lines. The background is a fantastical blend of European and Japanese elements, with floating symbols and dreamlike imagery. The overall effect is a magical and imaginative depiction, capturing Miró's surrealist style. --ar 2:3 --style raw --v 6.0

06. Piet Mondrian
皮特‧蒙德里安

抽象幾何構圖，強調垂直與水平線條的秩序美。

A detailed image in the style of Piet Mondrian, depicting a European medieval prince and a beautiful Japanese geisha. The scene uses Mondrian's iconic grid and primary color palette, with the prince and geisha represented through geometric shapes and intersecting lines. The prince is adorned in medieval attire with abstract, rectangular forms, while the geisha is depicted with simplified, colorful blocks representing her kimono. The background features a blend of European and Japanese architectural elements, arranged in a harmonious grid. The overall style is minimalist and abstract, reflecting Mondrian's De Stijl movement. --ar 2:3 --style raw --v 6.0

07. Paul Klee
保羅‧克利

幻想與童趣的結合，創造出詩意與符號的視覺語言。

A detailed image in the style of Paul Klee, depicting a European medieval prince and a beautiful Japanese geisha. The composition uses Klee's whimsical and symbolic style, with a focus on color, texture, and playful abstraction. The prince and geisha are rendered with intricate patterns and vibrant colors, with a blend of geometric and organic forms. The background is a fantastical landscape that combines European and Japanese elements in a dreamlike manner. The overall effect is a vibrant and imaginative depiction, capturing Klee's unique approach to art. --ar 2:3 --style raw --v 6.0

08. Kazimir Malevich
卡齊米爾‧馬列維奇

至上主義創始人，以簡化幾何形構築精神世界。

A detailed image in the style of Kazimir Malevich, depicting a European medieval prince and a beautiful Japanese geisha. The composition uses Malevich's Suprematist principles, focusing on basic geometric shapes and a limited color palette. The prince and geisha are represented through simplified, abstract forms, with the prince's medieval attire and the geisha's kimono rendered in bold, flat colors. The background features a minimalist blend of European and Japanese architectural elements, emphasizing pure form and color. The overall style is abstract and geometric, reflecting Malevich's avant-garde approach. --ar 2:3 --style raw --v 6.0

09. Alexander Calder
亞歷山大・考爾德

動態雕塑創始人，以輕盈優雅的移動雕塑著稱。

A detailed image in the style of Alexander Calder, depicting a European medieval prince and a beautiful Japanese geisha. The scene is characterized by Calder's use of playful, kinetic shapes and vibrant colors. The prince and geisha are rendered with simple, abstract forms, and bold, fluid lines. The background features a whimsical blend of European and Japanese elements, with mobile-like structures and dynamic compositions. The overall effect is a lively and imaginative depiction, capturing Calder's unique style. --ar 2:3 --style raw --v 6.0

10. Amedeo Modigliani
阿梅迪奧・莫迪利亞尼

擅長細長人物肖像，融合古典與現代風格。

A detailed image in the style of Amedeo Modigliani, depicting a European medieval prince and a beautiful Japanese geisha. The composition uses Modigliani's signature elongated forms and soft, muted colors. The prince and geisha are rendered with graceful, elongated features, and a sense of serene elegance. The background features a blend of European and Japanese architectural elements, with a focus on simplicity and harmony. The overall style is refined and expressive, capturing Modigliani's unique approach to portraiture. --ar 2:3 --style raw --v 6.0

當代藝術

一些在 1950 年之後活躍的當代藝術家，也是在 Midjourney 有風格標記的藝術家，包含他們的風格介紹。這一部份也引起一些團體關注，近期有關 Midjourney、Stable Diffusion 等平台未經授權使用藝術家風格的爭議引起了廣泛關注，特別是在法律方面。這些爭議主要涉及 AI 生成的藝術是否侵犯了藝術家的版權，以及這些 AI 模型如何使用受保護的作品來進行訓練。

● 爭議背景

多位藝術家，包括 Sarah Andersen、Kelly McKernan 和 Karla Ortiz，對 Midjourney 和 Stability AI 等公司提起了集體訴訟，指控這些公司在未經許可的情況下使用他們的作品進行 AI 模型的訓練。這些藝術家聲稱，AI 生成的圖像在風格上模仿了他們的作品，從而侵犯了他們的版權。

● 法律案件和判決

在訴訟過程中，位於舊金山的聯邦地區法官 William Orrick 駁回了針對 DeviantArt 和 Midjourney 的部分訴訟，理由是原告未能註冊他們的作品，因此不具備提出版權侵權的資格。然而，針對 Stability AI 的直接版權侵權指控仍被保留，法庭允許原告對其訴訟進行修正和補充。

Stability AI 和其他被告如 DeviantArt 則爭辯稱，他們僅僅是使用了其他公司開發的 AI 生成器，並沒有直接控制這些 AI 模型的製作過程。因此，他們認為，將其使用 AI 模型的行為視為侵權將對編程和媒體領域產生深遠且不合理的影響。

此外，AI 公司 Runway 指出，他們沒有存儲任何原始圖像，且 **AI 生成圖像並不是對受保護作品的精確複製**，而是基於風格和概念的創作。這些觀點強調了風格本身並不受版權保護的法律原則。

● 影響和未來展望

這些案件的裁決和進展對於 AI 生成藝術的法律框架具有深遠的影響。它們不僅涉及到藝術家的版權保護，也涉及到 AI 技術在創作領域的合法性和道德問題。這些案件的結果將影響未來 AI 技術在藝術創作中的應用和規範，並可能導致新的法律和政策的出臺，以更好地平衡創新技術和藝術家權益之間的關係。

這些爭議和法律案件顯示了 AI 技術在藝術創作中的潛力和挑戰，並強調了在使用這些技術時需要謹慎考慮的法律和道德問題。我個人則認為 AI 影像生成平台對於作品及風格進行訓練，相當於一個學習的過程，至於與創作者的權利義務關係，我們仍是期待風格的發明者能夠得到相對應的保障，但就實務上來看，所謂藝術風格屬於一個相對模糊的概念，且藝術創作本就是一個不斷傳承、演進的過程，例如常玉的作品，很明顯的有馬蒂斯的創作風格。在智財權的邏輯上，只要不是 100% 對於藝術品進行複製，或是冒稱藝術家創作，損及著作財產權及人格權，模擬、仿效、臨摹或是最近流行的向某某大師致敬這樣的說法，在法律上還是保護重作者的創作自由。但無論如何，尊重原創是一個創作者的道德倫理問題，我們可以借鏡學習，但不能占為己有。

01. Andy Warhol │ 安迪‧沃荷

波普藝術，以大膽的色彩和重複圖像聞名。

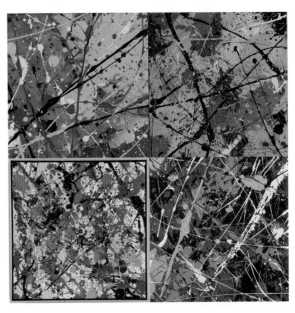

02. Jackson Pollock │ 傑克遜‧波洛克

抽象表現主義，著名的滴畫技法。

03. David Hockney │ 大衛‧霍克尼

英國流行藝術，色彩鮮豔且富有活力。

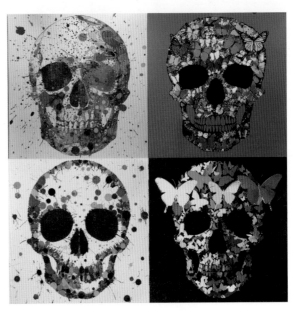

04. Damien Hirst │ 達米恩‧赫斯特

概念藝術，常使用動物標本和醫學相關材料。

05. Takashi Murakami │村上隆

│ 超扁平藝術，結合日本傳統和流行文化元素。

06. Jeff Koons │傑夫・昆斯

│ 波普藝術，以巨大且光滑的雕塑聞名。

07. Jenny Holzer │珍妮・霍爾澤

│ 概念藝術，常使用文本作為主要表現手段。

08. Yousuf Karsh │尤瑟夫・卡什

│ 攝影，以黑白人像攝影著名。

09. Barbara Kruger │ 芭芭拉 · 克魯格

| 概念藝術，結合文字和影像以傳達社會評論。

10. Chuck Close │ 塔克 · 費茲德

| 超寫實主義，以大型肖像畫聞名。

11. Keith Haring │ 基斯 · 哈林

| 塗鴉藝術，簡單且富有衝擊力的線條。

12. Jean-Michel Basquiat │ 尚 - 米榭 · 巴斯奇亞

| 塗鴉藝術，結合原始主義和表現主義。

13. Cindy Sherman ｜ 辛蒂‧舍曼
｜ 攝影，通過自我肖像探討身份和角色。

14. Richard Prince ｜ 理查‧普林斯
｜ 概念藝術，以重構和挪用影像聞名。

15. Ai Weiwei ｜ 艾未未
｜ 裝置藝術和概念藝術，常涉及政治和社會議題。

16. Cecily Brown ｜ 塞西莉‧布朗
｜ 表現主義，結合抽象和具象元素。

17. Mark Bradford ｜馬克‧布拉德福德
｜ 抽象藝術，作品常以城市地圖為靈感。

18. Anselm Kiefer ｜安瑟姆‧基弗
｜ 新表現主義，探討歷史和神話題材。

19. Robert Rauschenberg ｜羅伯特‧勞森伯格
｜ 新達達主義，結合繪畫、雕塑和裝置藝術。

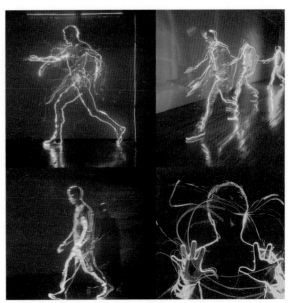

20. Bruce Nauman ｜布魯斯‧瑪瑟斯
｜ 概念藝術和裝置藝術，探討語言和身體。

21. Doris Salcedo ｜ 多麗斯 · 薩爾塞多

　　裝置藝術，探討暴力和記憶。

22. Rachel Whiteread ｜ 瑞秋 · 懷特瑞德

　　雕塑，常以日常物品為主題。

23. Kerry James Marshall
　　凱瑞 · 詹姆斯 · 馬歇爾

　　具象繪畫，探索非裔美國人歷史和文化。

24. Gerhard Richter ｜ 格哈德 · 裡希特

　　抽象和寫實繪畫，以模糊效果聞名。

25. Bill Viola ｜比爾‧維奧拉
録像藝術，探討人類情感和靈性。

26. Yayoi Kusama ｜草間彌生
裝置藝術和抽象畫，以波點和鏡像房間聞名。

27. Olafur Eliasson ｜奧拉維爾‧埃利亞松
裝置藝術，結合自然元素和科技。

28. Anish Kapoor ｜阿尼什‧卡普爾
雕塑，以大規模公共藝術作品著名。

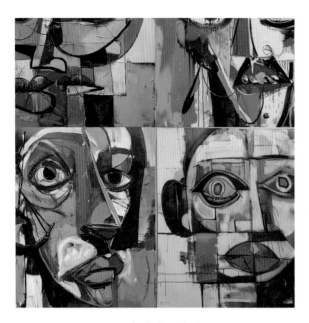

29. George Condo │喬治‧康多
│ 現實和超現實混合，具象與抽象並存。

30. Thomas Schütte │托馬斯‧沙羅羅克
│ 雕塑和裝置藝術，以奇異和反常的形式聞名。

以上我們附上以藝術家為名（Tokens），由 AI 根據其風格自行生成的影像，然而創作並非是一味的模擬仿效，我們應該善用 AI 對於藝術風格的定義，創造屬於自己的題材以及獨創性的內容。

我們以尚 - 米榭‧巴斯奇亞的風格，創作一幅釋迦牟尼佛，這種東方文化主題，以西方藝術形式表現，無庸置疑就是原創，這才是利用 AI 生成影像最大的意義。

生成範例│尚 - 米榭‧巴斯奇亞風之釋迦牟尼佛

A vibrant and chaotic illustration of Siddhartha Gautama (Buddha), drawn in the style of Jean-Michel Basquiat. The Buddha is depicted in a meditative pose, surrounded by Basquiat's signature graffiti elements, bold lines, and expressive brushstrokes. The composition includes a mix of symbols, crowns, and abstract shapes, with a palette of bright, contrasting colors. The overall scene captures the spiritual essence of Buddha, infused with the raw energy and intensity characteristic of Basquiat's art. --ar 2:3 --style raw --v 6.0

以尚 - 米榭‧**巴斯奇亞** (Jean-Michel Basquiat) **風格**繪製的喬達摩悉達多 (Siddhartha Gautama)（佛陀）充滿活力且混亂的插圖。佛陀以冥想的姿勢描繪，周圍環繞著巴斯奎特標誌性的塗鴉元素、大膽的線條和富有表現力的筆觸。構圖包括符號、皇冠和抽象形狀的混合，以及明亮、對比色的調色板。整個場景捕捉了佛陀的精神本質，注入了巴斯奎特藝術的原始能量和強度特徵。 --ar 2:3 -- 風格原始

各式水彩畫風格是我個人最喜歡應用的，有幾個主要的原因，一是水彩畫的容錯率高，二是在作品輸出後，水彩畫形式是除了攝影風格外最能貼近手繪原創作品的媒材。所謂容錯率就是目前 AI 生成的結果，由於畫素的限制，在許多畫面細節容易出現瑕疵，例如微小的臉部表情等等，但由於水彩畫的先天創作特性，本就不追求細緻的寫實，所以就更能模擬或接近手繪風格。

這裡還是要強調，套用藝術家 Token，不是要去剽竊藝術家的成果，而是基於目前 AI 影像模型的訓練原理，以藝術家為名的 Token，就含括著一幅作品大部分的技法及視覺效果。但所謂風格，也會受主題限制，不是每個主題套用任意風格都可以得到完美的效果。比較好的作法還是針對主題，挑選適合的風格帶入。

> 提示詞：Watercolors by {{ 畫家名 }}

01. Jean Haines ｜吉恩・海恩斯

｜英國，現代水彩畫家，以自由奔放的技法和鮮明的色彩聞名。

02. Thomas Schaller ｜托馬斯・沙勒

｜美國，以城市景觀和建築題材聞名，風格寫實且細膩。

03. Joseph Zbukvic｜約瑟夫・祖布科維奇

澳大利亞，以風景畫和都市風景著稱，技巧高超且富有氛圍感。

04. Alvaro Castagnet｜阿爾瓦羅・卡斯塔涅

烏拉圭，以強烈的光影對比和活力十足的筆觸聞名。

05. Dean Mitchell｜迪恩・米切爾

美國，以人像和社會主題的水彩作品著稱，風格細膩而富有情感。

06. Dong Kingman｜董錦根

美國，以城市景觀和風景畫聞名，色彩鮮明且具動感。

07. Mary Whyte ｜瑪麗‧懷特
｜ 美國，以現實主義的人像畫聞名，細膩且充滿情感。

08. Phil Dike ｜菲爾‧戴克
｜ 美國，現代水彩畫家，以鮮明的色彩和抽象表現聞名。

09. Rex Brandt ｜雷克斯‧布蘭特
｜ 美國，以海洋和風景畫著稱，風格自由且充滿活力。

10. Sir William Russell Flint
威廉‧羅素‧弗林特爵士
｜ 英國，以其優雅的女性人物畫著稱，風格浪漫而細膩。

11. Rowland Hilder｜羅蘭‧希爾德

英國，以英國鄉村景觀畫著稱，風格寫實且細緻。

12. Edward Wesson｜愛德華‧韋森

英國，以簡潔且具表現力的水彩風景畫聞名。

13. Elizabeth Murray｜伊麗莎白‧穆雷

美國，以抽象表現主義聞名，作品充滿動感和色彩。

14. John Sell Cotman｜約翰‧賽爾‧科特曼

英國，以風景畫著稱，風格柔和而細膩。

15. Beatrix Potter ｜ 碧雅翠絲 · 波特

｜ 英國，以其童書插畫和自然題材的水彩畫著稱。

16. John White Alexander
約翰 · 懷特 · 亞歷山大

｜ 美國，以優雅的人物畫和裝飾風格著稱。

17. James McNeill Whistler
詹姆斯 · 麥克尼爾 · 惠斯勒

｜ 美國，以其色調主義和細膩的畫風聞名。

18. Thomas Moran ｜ 托馬斯 · 莫蘭

｜ 美國，以壯麗的風景畫著稱，特別是美國西部的景觀。

19. Maurice Prendergast
　　莫里斯・普倫德加斯特
　　美國，後印象派畫家，色彩鮮豔的城市和海灘場景聞名。

20. Arthur Rackham ｜ 亞瑟・拉克姆
　　英國，以其童話插畫和幻想題材的水彩畫著稱。

21. Childe Hassam ｜ 查爾德・哈桑
　　美國，印象派畫家，以其都市和風景畫聞名。

22. Edward Hopper ｜ 愛德華・霍普
　　美國，以其孤獨的都市和鄉村場景聞名，風格寫實且富有
　　情感。

23. John Constable ｜約翰‧康斯特勞

英國，以風景畫聞名，特別是英國鄉村景觀。

24. Georgia O'Keeffe ｜喬治亞‧歐姬芙

美國，以其大型花卉和新墨西哥風景畫著稱，風格獨特且充滿力量。

25. Andrew Wyeth ｜安德魯‧懷斯

美國，以寫實風格和細膩的人物及風景畫著稱。

26. Winslow Homer ｜溫斯洛‧霍默

美國，以其海洋和戰爭題材的水彩畫著稱，風格真實且富有戲劇性。

27. John Singer Sargent
約翰・辛格・薩金特

美國，以其優雅的人物畫和風景畫聞名，風格細膩且充滿活力。

28. J.M.W. Turner | **威廉・透納**

英國，以其浪漫主義風景畫和劇烈的光影效果聞名。

29. Milford Zornes | **米爾福德・佐恩斯**

美國，以其鮮明的色彩和抽象風景畫著稱。

30. Richard Schmid | **理查・施密德**

美國，以寫實主義的風景和人像畫著稱，技法精湛且細膩。

在 Midjourney 中我們可以利用攝影家的藝術風格，生成具有特定風格的影像，首先我們必須瞭解攝影家的藝術風格，這些攝影風格可能包括特殊的攝影技巧、構圖、主題等等，在使用特定風格時，請務必注意主題與該風格的適配性，才能得到最佳的生成效果，例如我們要生成一張風景圖，卻使用社會紀實攝影師的風格，可能格格不入，但也可能會有意想不到的效果。

Sunrise photography of the Matterhorn, using the style of Dorothea Lange
馬特洪峰的日出攝影，使用 Dorothea Lange 的風格。

Taipei street protests, styled by Dorothea Lange.
台北街頭抗議活動，則可能更貼近紀實攝影師 Dorothea Lange 的創作風格。

我們還是要不斷的提醒讀者，使用 AI 生成創作的目的不是直接複製某個作品，而是利用 AI 建立的模型風格，生成屬於自己獨一無二的主題內容，主題與風格的搭配，可以協調一致，也可以是衝突矛盾的，最後呈現出的視覺效果有賴於大量的實驗生成。以下我們列舉了 100 位世界知名的攝影風格，並以 AI 進行生成供大家進行日後創作參考，非攝影師原創作品。

01. Ansel Adams │ 安塞爾・亞當斯
│ 美國，以黑白風景攝影聞名，擅長捕捉自然風光。

**02. Henri Cartier-Bresson
亨利・卡蒂埃 - 布列松**
│ 法國，街頭攝影的先驅，捕捉決定性瞬間。

03. Robert Capa │ 羅伯特・卡帕
│ 匈牙利，戰地攝影師，記錄戰爭的殘酷。

04. Gordon Parks │ 戈登・帕克斯
│ 美國，記錄社會不公的攝影師，擅長人文紀實。

05. Dorothea Lange ｜桃樂西婭 · 蘭格

｜ 美國，以大蕭條時期的紀實攝影聞名。

06. Richard Avedon ｜理查 · 阿維頓

｜ 美國，時尚和肖像攝影大師，風格獨特。

07. Steve McCurry ｜史蒂夫 · 麥柯裡

｜ 美國，捕捉全球文化和衝突的攝影師，作品《阿富汗少女》
｜ 聞名。

08. Irving Penn ｜歐文 · 佩恩

｜ 美國，時尚攝影大師，以其簡潔的構圖和優雅的風格著稱。

09. Robert Frank ｜羅伯特 · 弗蘭克

｜ 瑞士 - 美國，《美國人》攝影集展現了美國社會的真實面貌。

10. Annie Leibovitz ｜安妮 · 萊博維茨

｜ 美國，時尚和名人肖像攝影大師。

11. Gordon Willis ｜戈登 · 威利斯

｜ 美國，電影攝影師，以其在《教父》三部曲中的傑出作品
｜ 著稱。

12. Daniel Errázuriz ｜丹尼爾 · 埃爾

｜ 智利，概念攝影師，作品融合藝術和社會批評。

13. Arnold Newman ｜阿諾爾德‧紐曼
｜ 美國，環境肖像攝影的開創者。

14. Elliott Erwitt ｜艾略特‧厄威特
｜ 法國 - 美國，以幽默和人性的捕捉著稱。

15. Sebastião Salgado ｜賽巴斯提奧‧薩爾加多
｜ 巴西，人文和紀實攝影大師，關注社會和環境問題。

16. Walker Evans ｜沃克‧埃文斯
｜ 美國，以記錄美國南部的貧困生活聞名。

17. André Kertész │ 安德列・柯特茲

│　匈牙利 - 美國，先鋒派攝影師，以其創新的構圖著稱。

18. Ersatz Voss │ 埃爾斯・沃斯

│　德國，實驗攝影師，以其獨特的視覺效果聞名。

19. Bill Brandt │ 比爾・布蘭特

│　英國，超現實主義攝影師，以黑白作品著稱。

20. Inge Morath │ 英格・莫拉斯

│　奧地利 - 美國，瑪格南圖片社成員，紀錄和肖像攝影師。

21. Martin Parr ｜馬丁‧帕爾
｜ 英國，色彩豐富、風格鮮明的紀實攝影師。

22. Lisa Kristine ｜麗莎‧克裡斯汀
｜ 美國，人權和人道主義攝影師，作品關注全球奴役現象。

23. Albert Watson ｜阿爾伯特‧沃森
｜ 蘇格蘭，時尚和名人肖像攝影大師。

24. Mary Ellen Mark ｜瑪麗‧艾倫‧馬克
｜ 美國，紀實攝影師，關注邊緣人物和社會問題。

25. Peter Lindbergh ｜彼得・林德伯格
｜ 德國，時尚攝影大師，以黑白攝影聞名。

26. Spencer Tunick ｜斯賓塞・圖尼克
｜ 美國，以大規模裸體人像攝影聞名。

27. Maggie Taylor ｜瑪姬・泰勒
｜ 美國，超現實主義數字攝影師。

28. Dnis Darzacq ｜德尼・達爾紮克
｜ 法國，紀實攝影師，關注都市生活。

29. Jan de Graaff ｜賈梅・德拉斯
｜ 荷蘭，紀實攝影師，擅長捕捉日常生活中的瞬間。

30. Dana Gluckstein ｜德納・吉爾福德
｜ 美國，人權和社會正義攝影師。

31. Josef Koudelka ｜約瑟夫・庫德爾卡
｜ 捷克-法國，以流浪者系列和記錄布拉格之春聞名。

32. Marc Riboud ｜馬克・魯本斯坦
｜ 法國，記錄中國文化和社會變遷的攝影師。

33. Harry Callahan ｜哈裡 · 卡拉漢

｜ 美國，以創新的城市風景和抽象攝影聞名。

34. Bruno Barbey ｜布魯諾 · 巴貝

｜ 摩洛哥 - 法國，紀實攝影師，記錄了全球多地的文化和社會變遷。

35. William Eggleston ｜威廉 · 埃格爾斯頓

｜ 美國，彩色攝影的先鋒，以捕捉日常生活中的美聞名。

36. Andre Lotter ｜安德列 · 洛特

｜ 德國，建築和城市風景攝影師。

37. Raymond Depardon ｜雷蒙·德波登

｜ 法國，瑪格南圖片社成員，記錄全球衝突和社會問題。

38. W. Eugene Smith ｜尤金·史密斯

｜ 美國，紀錄和戰地攝影師，作品《水俁》聞名。

39. Bob Gruen ｜鮑勃·格魯恩

｜ 美國，搖滾攝影師，以記錄音樂明星聞名。

40. Francesca Woodman ｜法蘭西斯·費斯

｜ 美國，超現實主義自畫像攝影師。

41. Alfred Eisenstaedt
　　阿爾弗雷德・艾森斯塔特
　| 德國 - 美國，記錄 20 世紀的重大歷史事件。

42. Lee Miller ｜ 李・米勒
　| 美國，戰地攝影師和時尚攝影師。

43. Helmut Newton ｜ 赫爾穆特・紐頓
　| 德國 - 澳大利亞，以大膽的時尚攝影聞名。

44. Lillian Bassman ｜ 莉利安・巴斯曼
　| 美國，時尚攝影師，以抽象風格聞名。

45. Edward Weston │ 愛德華 · 韋斯頓

| 美國，靜物和人體攝影大師。

46. Herman Leonard │ 赫爾曼 · 斯科特

| 美國，爵士攝影師，記錄了大量的音樂歷史。

47. Mark Seliger │ 馬克 · 塞爾格

| 美國，名人肖像攝影師，以獨特的風格聞名。

48. Ian Marcus │ 伊恩 · 馬庫斯

| 英國，時尚和商業攝影師。

49. Alvin Langdon Coburn ｜阿爾文・蘭格多

｜ 英國，美國，先鋒派攝影師。

50. Hiroshi Sugimoto ｜杉本博司

｜ 日本，以長時間曝光拍攝的劇院和海景照片著名，捕捉時間的流逝。

51. Joseph Markus ｜約瑟夫・馬庫斯

｜ 美國，紀錄和肖像攝影師。

52. Mary Voss ｜瑪麗・沃斯

｜ 德國，實驗攝影師，以獨特的視覺效果聞名。

53. John Dawson ｜約翰·唐森
｜ 美國，紀實和風景攝影師。

54. Barbara Kruger ｜芭芭拉·克魯格
｜ 美國，概念攝影師，擅長文字與圖像結合。

55. David LaChapelle ｜大衛·拉查佩爾
｜ 美國，時尚和廣告攝影師，以超現實和戲劇性風格著稱。

56. Garry Winogrand ｜格裡·溫諾格蘭
｜ 美國，街頭攝影大師，記錄了 20 世紀美國的日常生活。

57. Nick Knight ｜尼克‧奈特
｜英國，時尚攝影師，以創新和實驗風格聞名。

58. Philip-Lorca diCorcia
　　菲力浦‧洛卡‧迪科西亞
｜美國，擅長在日常場景中捕捉戲劇性瞬間。

59. Lawrence Carroll ｜勞倫斯‧卡羅爾
｜美國，風景和自然攝影師。

60. William Klein ｜威廉‧克萊因
｜法國，美國，街頭攝影師和時尚攝影師。

61. Matthew Pillsbury ｜ 馬修・皮爾斯伯裡

｜ 美國，風景和詳細自然攝影。

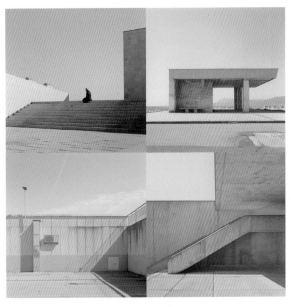

62. Matthias Heiderich ｜ 馬蒂亞斯・海德里希

｜ 德國，抽象和極簡主義風格。

63. Mert Alas ｜ 梅特・阿拉斯

｜ 土耳其，時尚和人像攝影。

64. Mitch Dobrowner ｜ 米奇・多布羅納

｜ 美國，風暴和自然攝影。

65. Moises Levy ｜摩西‧利維

｜ 墨西哥，黑白建築和風景攝影。

66. Nirav Patel ｜尼拉夫‧派特爾

｜ 美國，情感和氛圍人像攝影。

67. Nona Faustine ｜諾娜‧福斯汀

｜ 美國，歷史和社會正義題材。

68. Patricia Voulgaris ｜派特裡夏‧沃爾加里斯

｜ 美國，幾何和黑白人像攝影。

69. Alex Prager │艾力克斯 · 普拉格
│ 美國，以其戲劇性和色彩豐富的敘事性照片聞名。

70. Sanne Sannes │桑內 · 桑尼斯
│ 荷蘭，情感和暗黑風格人像攝影。

71. Shomei Tomatsu │富士宏
│ 日本，戰後社會和文化攝影。

72. Terence Donovan │特倫斯 · 多諾萬
│ 英國，經典黑白人像攝影。

73. James Nizam ｜ **詹姆斯 · 尼贊**

｜ 加拿大，概念和實驗攝影。

74. Jerry N. Uelsmann ｜ **傑裡 · 尤爾斯曼**

｜ 美國，超現實主義和多重曝光攝影。

75. Jessica Backhaus ｜ **潔西嘉 · 巴克豪斯**

｜ 德國，靜物和情感攝影。

76. Joel Peter Witkin ｜ **喬爾 · 彼得 · 維特金**

｜ 美國，超現實主義和黑暗風格攝影。

77. Petra Collins ｜ 彼得・科林斯
｜ 加拿大，夢幻和彩色人像攝影。

78. Claude Cahun ｜ 克勞德・卡洪
｜ 法國，超現實主義和 LGBTQ+ 主題攝影。

79. Neil Krug ｜ 尼爾・克魯格
｜ 美國，夢幻和色彩豐富的攝影風格。

80. Kathy Fornal ｜ 凱西・福納爾
｜ 美國，花卉和夢幻風格攝影。

81. Edward Burtynsky ｜愛德華・伯汀斯基

｜ 加拿大，以紀錄工業景觀和環境問題的攝影聞名。

82. Zanele Muholi ｜紮內勒・穆霍利

｜ 南非，以 LGBTQ+ 社群和性別議題的肖像攝影聞名。

83. Richard Mosse ｜理查・莫斯

｜ 愛爾蘭，以使用紅外攝影技術捕捉衝突地區的影像聞名。

84. André Kertész ｜安德列・凱爾特斯

｜ 匈牙利 - 美國，先鋒派攝影師，以其創新的構圖著稱。

85. Robert Mapplethorpe
羅伯特・馬普爾索普

美國，黑白肖像和花卉攝影。

86. Jeff Wall | **傑夫・沃爾**

加拿大，戲劇性和概念攝影。

87. Lee Friedlander | **李・弗裡德蘭德**

美國，以紀錄美國社會和文化的黑白攝影著稱。

88. Alec Soth | **亞歷克・索斯**

美國，以大幅紀實攝影作品聞名，探索美國人的生活。

89. Andreas Gursky │ 安德莉亞斯·古爾斯基
│ 德國，大尺度的高解析度攝影。

90. Joel Meyerowitz │ 喬爾·邁耶羅威茨
│ 美國，紀實和街頭攝影師，以彩色作品聞名。

91. Saul Leiter │ 索爾·萊特
│ 美國，以彩色街頭攝影和抽象構圖著稱。

92. Herb Ritts │ 赫伯·瑞茲
│ 美國，時尚和名人肖像攝影師。

93. Daido Moriyama ｜森山大道

｜ 日本，以黑白街頭攝影聞名，風格粗獷且具實驗性。

94. Brassaï ｜布拉塞

｜ 法國，紀實攝影師，以捕捉巴黎夜生活著稱。

95. Nan Goldin ｜南‧戈丁

｜ 美國，紀實攝影師，以紀錄 1980 年代的紐約地下文化著稱。

96. Martin Parr ｜馬丁‧帕爾

｜ 英國，紀實攝影師，以捕捉日常生活的幽默和荒誕著稱。

97. Vivian Maier ｜薇薇安‧邁爾
｜ 美國，街頭攝影師，後期才被發現其大量紀實作品。

98. Diane Arbus ｜黛安‧阿勃斯
｜ 美國，以紀錄社會邊緣人物和怪異主題聞名。

99. Cindy Sherman ｜辛蒂‧舍曼
｜ 美國，通過自我扮演和自我肖像來探索身份和社會角色。

100. Gregory Crewdson ｜葛列格里‧克魯德森
｜ 美國，以戲劇性且超現實的場景聞名，創作如電影場景般的
｜ 攝影作品。

如果對於各式空間建築風格有基本的認識，日後在創作各類影像畫面時，就能夠更輕易掌握畫面中建築物的型態，使得畫面更精緻，更準確地達到創作者的定義及要求。例如，多數人在寫提示詞時，普遍只會寫一間房子、一棟大廈之類的提示詞，如果能夠充分瞭解各類建築風格，就可以寫成一間巴洛克風格建築的房子、安藤忠雄建築風格的大樓。以下我們分別以歷史、區域文化以及建築師將建築風格進行分類，並提供生成影像供讀者參考，在生成影像時，將不同格的建築物搭配一些不容易出現的場景，例如極地、山頂等等，而這也是 AI 影像生成有趣之處，只要你想像力到哪，創作就到哪。

此外，這單元所有的生成影像提示詞，媒材均是以寫實照片呈現，讀者可嘗試置換為前面所介紹的各類媒材，例如素描、雕塑等等。

依時間分類

01. Classical Greek Architecture
古典希臘建

特點 ⇒ 柱式、對稱、比例。

主要時期 ⇒ 西元前 5 世紀至西元前 3 世紀。

生成主題｜沙漠中的希臘建築

A photorealistic image depicting a classical Greek building set in a desert landscape. The structure features iconic Greek architectural elements such as tall Doric columns, marble statues, and intricate friezes. The building stands majestically amidst the vast, sandy desert, with dunes stretching out in the background. The sunlight casts dramatic shadows, highlighting the details of the ancient architecture. The atmosphere is serene and surreal, with a clear blue sky and a few wisps of clouds. The overall style is highly detailed, capturing the timeless elegance of Greek architecture in the stark, contrasting environment of the desert. --ar 3:2 --style raw

2

3

4

5

02. Ancient Roman Architecture
古羅馬建築

特點 ⇒ 拱門、圓頂、混凝土。

主要時期 ⇒ 西元前 1 世紀至西元 5 世紀。

生成主題｜綠洲中的古羅馬建築

A photorealistic image depicting an ancient Roman building set in a desert oasis. The structure features iconic Roman architectural elements such as tall Corinthian columns, grand arches, and intricate frescoes. The building stands majestically amidst a lush green oasis, surrounded by palm trees and a clear blue water source. The contrast between the arid desert landscape and the verdant oasis creates a striking and serene scene. The sunlight casts warm, golden hues, highlighting the details of the Roman architecture and the vibrant greenery. The overall style is highly detailed, capturing the timeless elegance of Roman architecture in the unique and refreshing setting of a desert oasis. --ar 3:2 --style raw

03. Gothic Architecture
哥德式建築

特點 ⇒ 尖頂拱門、飛扶壁、大窗戶和彩色玻璃。

主要時期 ⇒ 12 世紀至 16 世紀。

生成主題｜阿爾卑斯山頂的哥德式建築

A photorealistic image depicting a Gothic-style building perched atop a peak in the Alps. The building stands majestically against the backdrop of the snow-capped Alps, with the rugged mountains and clear blue sky creating a dramatic and awe-inspiring scene. The atmosphere is serene and majestic, with the sunlight highlighting the details of the Gothic architecture and the pristine beauty of the alpine landscape. The overall style is highly detailed, capturing the timeless elegance of Gothic architecture in the stunning natural setting of the Alps. --ar 3:2 --style raw

04. Renaissance Architecture
文藝復興建築

特點 ⇒ 對稱、比例、古典柱式的復興。

主要時期 ⇒ 14 世紀至 17 世紀。

生成主題｜在北極的文藝復興風格建築

A photorealistic image depicting a Renaissance-style building set in the Arctic landscape. The structure features iconic Renaissance architectural elements such as domes, symmetrical facades, arches, and intricate sculptures. The building stands majestically amidst the vast, icy expanse of the Arctic, with snow and ice surrounding it. The stark contrast between the classical elegance of the Renaissance architecture and the pristine, harsh Arctic environment creates a surreal and captivating scene. The background includes snow-capped mountains and a clear blue sky, with the cold sunlight casting a soft glow on the building's details. The overall style is highly detailed, capturing the timeless beauty of Renaissance architecture in the unique and stark setting of the Arctic. --ar 3:2 --style raw

05. Baroque Style
巴洛克風格

特點：華麗、動感、複雜的裝飾、豐富的色彩。

主要時期 ⇒ 17 世紀至 18 世紀初。

生成主題｜海上的巴洛克建築

A photorealistic image depicting a Baroque-style castle built on the sea, set in a futuristic and surreal environment. The castle features intricate architectural details, ornate sculptures, and grandiose towers, all rendered in the opulent Baroque style. The scene is imbued with a sci-fi atmosphere, with advanced technology seamlessly integrated into the classical architecture. Strange, luminescent marine life can be seen swimming around the castle, casting an eerie glow. The sky above is filled with otherworldly colors and celestial phenomena, creating a sense of wonder and mystery. The overall style is highly detailed and imaginative, capturing the fusion of Baroque elegance and futuristic fantasy. --ar 3:2 --style raw

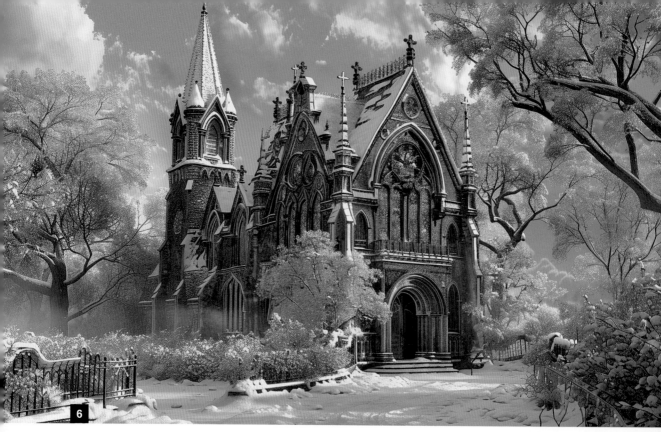

06. Victorian Architecture
維多利亞風格

特點 ⇒ 複雜的外立面裝飾、彩色玻璃、維多利亞女王時期風格。

主要時期 ⇒ 19 世紀中期至 19 世紀末。

生成主題｜雪地裡的維多利亞風格建築

A photorealistic image depicting a Victorian-style church set in a snowy landscape. The church features intricate gothic architectural details, tall spires, stained glass windows, and ornate stone carvings, all characteristic of the Victorian era. The scene is enveloped in a serene blanket of snow, with the soft white covering the ground and dusting the rooftops. The atmosphere is peaceful and quiet, with the glow from the church's windows casting a warm light on the surrounding snow. The background includes a clear winter sky with a few scattered clouds, enhancing the tranquil and majestic beauty of the snowy Victorian church. --ar 3:2 --style raw

07. Neoclassicism
新古典主義

特點 ⇒ 對稱、簡潔、典雅。

主要時期 ⇒ 18 世紀中期至 19 世紀初。

生成主題｜草原上的新古典主義建築

A photorealistic image depicting a Neoclassical-style building set in a vast, open prairie. The structure features iconic Neoclassical architectural elements such as tall Corinthian columns, symmetrical facades, and a grand pediment adorned with intricate sculptures. The building stands elegantly amidst the rolling grasslands, with wildflowers and tall grasses gently swaying in the breeze. The background includes a clear blue sky with a few scattered clouds, enhancing the serene and timeless beauty of the scene. The sunlight casts a warm glow, highlighting the details of the Neoclassical architecture and the vibrant greenery of the prairie. The overall style is highly detailed, capturing the grandeur and elegance of Neoclassical architecture in the peaceful and expansive setting of the grasslands. --ar 3:2 --style raw

08. Modernism
現代主義

特點 ⇒ 簡約、功能主義、使用現代材料。

主要時期 ⇒ 20 世紀初至 20 世紀中期。

生成主題｜現代主義建築

Modernism building --style raw

以下我們也可以單純指定建築物樣式，忽略背景、形制等描述。

09. Postmodernism
後現代主義

特點 ⇒ 混合多種風格、注重裝飾、對現代主義的反動。

主要時期 ⇒ 20 世紀 70 年代至 90 年代。

生成主題｜後現代主義風格建築

Postmodernism building --style raw

10. High-Tech Architecture
高科技建築

特點 ⇒ 暴露技術結構、使用高科技材料、強調工業美學。

主要時期 ⇒ 20 世紀 70 年代至現代。

生成主題｜高科技風格建築

High-Tech Architecture building --style raw

依區域分類

01. Traditional Chinese Architecture 中國傳統建築

特點 ⇒ 對稱佈局、木結構、飛簷翹角。

主要區域 ⇒ 中國。

生成主題 | 2300 年一棟保存良好的中國傳統建築

photorealistic image depicting, preserved traditional Chinese building set in the year 2300. The structure features iconic, upturned eaves, wooden carvings, and red lanterns, reflecting classic, building, futuristic surroundings, sleek —ar 3:2

02. Japanese Architecture 日本建築

特點 ⇒ 簡約、自然材料、和式庭園。

主要區域 ⇒ 日本。

生成主題 | 森林裡的傳統日本建築

photorealistic image depicting a traditional Japanese building nestled within a dense forest, warm glow of the setting sun. The structure features iconic, shoji screens, tatami mats, and a gently sloping, roof, forest --ar 3:2 --style raw

03. Islamic Architecture 伊斯蘭建築

特點 ⇒ 圓頂、拱門、幾何圖案。

主要區域 ⇒ 中東、北非。

生成主題｜冰原上的伊斯蘭建築

photorealistic image depicting an Islamic building set on a vast, icy plain. The structure, Islamic architectural elements such as domes, minarets, and intricate geometric patterns. The building, majestically, frozen, snow and ice, reflecting, blue --ar 3:2 --style raw

04. Scandinavian Architecture 斯堪地那維亞建築

特點 ⇒ 簡約、自然、功能性強。

主要區域 ⇒ 北歐。

生成主題｜丘陵地上斯堪的納維亞建築

photorealistic image depicting a Scandinavian-style building set in a rolling hillside landscape. The structure, iconic, Scandinavian architecture, such as clean lines, large windows, wooden facades, roof, blends harmoniously, grassy hill --ar 3:2

05. Mediterranean Architecture 地中海建築

特點 ⇒ 白色石膏牆、拱門、紅瓦屋頂。
主要區域 ⇒ 地中海沿岸。

生成主題 │ 懸崖邊上的地中海建築

photorealistic image depicting a Mediterranean-style building perched, cliff. The structure features iconic, Mediterranean, whitewashed, terracotta roof tiles, and arched windows. The building, majestically, backdrop, ocean, with waves crashing, rocky --ar 2:3

06. Tropical Architecture 熱帶建築

特點 ⇒ 採開放式設計、自然且通風、使用當地材料。
主要區域 ⇒ 東南亞、南美。

生成主題 │ 南極冰原上的熱帶建築

photorealistic image depicting a tropical-style building set on an icy plain in Antarctica. The structure features iconic, tropical architecture, such as open-air design, thatched roofs, and wooden beams, against, frozen, snow --ar 3:2

依知名建築師分類

前面章節我們已經介紹過不同時期以及區域上的建築風格表現，下麵我們將介紹幾位國際知名建築師的風格表現，理解並善用這些不同建築風格，都有助於豐富影像生成的細節。

01. Le Corbusier 勒·柯布西耶

特點 ⇒ 功能主義、五點建築原則。
代表作品 ⇒ 薩伏伊別墅、聯合國總部大樓。

生成主題 | 草原上的 Le Corbusier

photorealistic aerial view of a building designed in the style of Le Corbusier, set in a vast prairie landscape. The structure features iconic elements of Le Corbusier's, style, such as clean lines, geometric, flat roofs, and pilotis (supporting columns)

02. Frank Lloyd Wright 弗蘭克·洛伊德·賴特

特點 ⇒ 有機建築、與自然融合。
代表作品 ⇒ 流水別墅、古根海姆博物館。

生成主題 | 森林裡的 Frank Lloyd Wright

photorealistic image depicting a building designed in the style of Frank Lloyd Wright, seamlessly integrated into a lush forest landscape. The structure features iconic, Wright, style, such as horizontal lines, flat or low-pitched, overhangs, Large windows

03. Louis Kahn 路易・康

特點 ⇒ 使用磚石和混凝土、幾何形狀。
代表作品 ⇒ 印度國家議會大廈、金鐘獎博物館。

生成主題 | **城市公園裡的 Louis Kahn**

photorealistic image depicting a building designed in the style of Louis Kahn, set within an urban park. The structure features iconic elements of Kahn, monumental geometric, natural light, void spaces, 's façade, exposed

04. Zaha Hadid 紮哈・哈迪德

特點 ⇒ 流動曲線、不對稱設計。
代表作品 ⇒ 倫敦水上運動中心、廣州大劇院。

生成主題 | **未來城市景觀的 Zaha Hadid**

photorealistic image depicting a building designed in the style of Zaha Hadid, futuristic urban landscape. The structure features iconic elements of Hadid's architectural style, such as fluid, organic, curves, materials. The building, façade is characterized

05. Tadao Ando 安藤忠雄

特點 ⇒ 使用裸露混凝土、簡約設計。
代表作品 ⇒ 光之教堂、住吉的長屋。

生成主題 | **自然環境中的安藤忠雄**

photorealistic image depicting a building designed in the style of Tadao Ando, within a tranquil natural environment. The structure, iconic elements of Ando's, minimalist concrete walls, clean, natural light and shadow. The building, façade

06. Ludwig Mies van der Rohe 密斯‧凡‧德羅

| **特點** ⇒ 少即是多、簡約主義。
| **代表作品** ⇒ 巴塞羅那館、範斯沃斯之家。

生成主題│城市景觀中的 Ludwig Mies van der Rohe

photorealistic image depicting a building designed in the style of Ludwig Mies van der Rohe, set in a minimalist urban landscape. The structure, iconic, Mies van der Rohe's architectural, clean lines, open floor plans, large glass walls, concrete, characterized

07. Richard Rogers 理查‧羅傑斯

| **特點** ⇒ 高科技建築、暴露結構。
| **代表作品** ⇒ 龐畢度中心、勞合社大樓。

生成主題│城市中的 Richard Rogers

photorealistic image depicting a building designed in the style of Richard Rogers, set within a vibrant urban landscape. The structure features iconic elements of Rogers, tech modernism, bold colors, and extensive, glass and steel, façade

08. Norman Foster 諾曼‧福斯特

| **特點** ⇒ 高科技建築、可持續設計。
| **代表作品** ⇒ 千禧橋、倫敦市政廳。

生成主題│城市中的 Norman Foster

photorealistic image depicting a building designed in the style of Norman Foster, sophisticated urban landscape. The structure, iconic elements of Foster's architectural, sleek, lines, extensive, glass and steel, sustainability, building, façade, curved

09. Santiago Calatrava
聖地亞哥‧卡拉特拉瓦

特點 ⇒ 雕塑般的建築、動感結構。

代表作品 ⇒ 瓦倫西亞藝術科學城、紐約世貿中心交通樞紐。

生成主題｜依水而建的 Santiago Calatrava

photorealistic image depicting a building designed in the style of Santiago Calatrava, set within a dynamic urban waterfront landscape. The structure features iconic elements of Calatrava's, organic, flowing, use of materials, and a striking, façade

10. Peter Behrens 畢爾‧鮑亞斯

特點 ⇒ 工業建築設計、德意志製造聯盟。

代表作品 ⇒ AEG 渦輪機廠、霍赫斯大廈。

生成主題｜工業風的 Peter Behrens

photorealistic image depicting a building designed in the style of Peter Behrens, set within an industrial urban landscape. The structure features iconic elements of Behrens' architectural style, such as a focus on functionalism, geometric, industrial, brick, building

這些分類展示了建築和空間設計在不同時期、地區和建築師風格上的多樣性和豐富性，為創作和研究提供了廣泛的參考。

瞭解不同的動漫插畫風格，將有助於您在提示詞中更準確地描述所需的效果。常見的動漫風格包括：

● **日本漫畫風格**：線條清晰，角色具有大眼睛、小鼻子等特徵。

● **動畫電影風格**：如吉卜力工作室的作品，具有細膩的背景和獨特的色彩運用。

● **科幻風格**：融合未來主義元素，充滿機械和高科技感。

● **奇幻風格**：以魔法、傳說生物和奇幻世界為主題。

相較於前面介紹的繪畫風格與攝影風格，插畫及動漫風格對於主題的「寬容度」較高，這也符合現實的邏輯，例如吉卜力工作室的作品給人的印象是溫馨的城市或是鄉村題材，但我們也可以利用這風格去創作科幻故事。

生成範例 | 吉卜力風格的星際大戰場景

Star Wars scene in the style of Studio Ghibli, blending, whimsical atmosphere of Ghibli, universe, serene, mystical landscape like, Princess Mononoke, spaceships

吉卜力風格的星際大戰場景，融合了吉卜力的異想天開的氛圍，宇宙，寧靜，神秘的景觀，魔法公主，太空船。

接下來整理出許多歐、美、日插畫家和漫畫家，並以其風格生成影像，無設定主題，由 AI 根據藝術家風格生成，非藝術家原創作品。

20 位美國插畫家和漫畫家：

01. Chris Samnee ｜ 克裡斯・薩姆尼

以細膩的線條和柔和的色調聞名，風格樸實且充滿力量。

02. Chip Zdarsky ｜ 奇普・茲達斯基

以幽默和深刻的角色描寫著稱，作品充滿細緻和生動的線條，風格鮮明且充滿個性。

03. Dave Stevens ｜ 戴夫・斯蒂文斯

著名作品包括《火箭人》，以經典的復古風格和精細的線條著稱，作品充滿懷舊氛圍。

04. Andy Kubert ｜ 安迪・庫伯特

以流暢的動作場景和細膩的角色描寫著稱，風格柔和且充滿動感。

05. Peter Bagge │ 彼得・巴格

以漫畫風格的瘋狂和幽默著稱，作品充滿大膽的線條和獨特的角色設計。

06. Ryan Ottley │ 里安・奧特利

作品以細膩的線條和詳細的角色描寫聞名，風格鮮明且充滿活力。

07. Mark Brooks │ 馬克・布魯克斯

以精緻的角色描寫和鮮明的色彩運用著稱，作品風格詳細且富有情感。

08. Joe Jusko │ 喬・賈斯科

以幻想和科幻插畫著稱，風格細緻且充滿想像力，作品充滿視覺震撼力。

09. Jack Kirby｜傑克‧科比

以在漫畫界的開創性工作聞名，風格大膽且充滿創意，作品充滿動感和視覺衝擊力。

10. J. Scott Campbell｜J. 斯科特‧坎貝爾

以細膩的角色描寫和動作場景聞名，風格鮮明且充滿活力。

11. Greg Capullo｜葛列格‧卡普洛

以黑暗且富有情感的風格聞名，整體作品充滿動感和視覺衝擊力。

12. Tom Grummett｜湯姆‧格魯梅特

以詳細的角色描寫和幻想風格著稱，作品充滿生動的線條和視覺震撼力。

13. Abbey Lossing ｜阿比‧羅辛

擅長編輯和雜誌插畫，風格鮮明且富有現代感，作品充滿視覺吸引力。

14. James Avati ｜詹姆斯‧阿瓦提

以憂鬱的場景和詳細的畫風著稱，作品風格充滿情感和視覺深度。

15. James Thurber ｜詹姆斯‧瑟伯

以幽默的漫畫和獨特的畫風聞名，整體作品風格輕鬆且充滿趣味。

16. Julie Paschkis ｜朱莉‧帕斯基斯

以鮮明的色彩和細緻的線條著稱，作品風格充滿幻想和視覺吸引力。

17. Kate Leth │凱特‧雷斯

以動作場景和鮮明的角色描寫著稱,作品風格生動且充滿活力。

18. Jenny Frison │珍妮‧弗裡森

以詳細的角色描寫和柔和的色調聞名,作品風格充滿情感和視覺深度。

19. Brian Michael Bendis
布萊恩‧邁克爾‧本迪斯

以漫畫書創作和動作場景聞名,作品風格鮮明且充滿視覺衝擊力。

20. Ed Brubaker │艾德‧布魯貝克

以其黑暗、細緻和充滿情感的漫畫書風格聞名,作品充滿動作和視覺吸引力。

20 位歐洲插畫家和漫畫家

01. Adrienne Segur ｜阿德里安‧塞古爾

法國，以童書插畫和細緻的動物描寫聞名，作品風格柔和且充滿幻想。

02. Gustav-Adolf Mossa
古斯塔夫 - 阿道夫‧莫薩

法國，以超現實主義和細緻的畫風著稱，作品風格獨特且充滿視覺震撼力。

03. André Franquin ｜安德魯‧佛蘭克

比利時，以漫畫書創作和動作場景聞名，作品風格生動且充滿活力。

04. Chris Foss ｜克裡斯‧福斯

英國，以科幻插畫和詳細的場景描寫著稱，作品風格鮮明且充滿想像力。

05. Henry Justice Ford
亨利 · 賈斯特斯 · 福特

英國，以細緻的動物描寫和柔和的色調聞名，作品風格充滿視覺吸引力。

06. George Perez ｜喬治 · 佩雷斯

西班牙，以漫畫書創作和動作場景聞名，作品風格鮮明且充滿視覺衝擊力。

07. Joe Madureira ｜喬 · 馬杜雷拉

英國，以漫畫書創作和動作場景聞名，作品風格鮮明且充滿視覺衝擊力。

08. Allie Brosh ｜艾莉 · 布羅什

英國，以漫畫書創作和動作場景聞名，作品風格鮮明且充滿視覺衝擊力。

09. Cecily Brown ｜塞西莉・布朗

英國，以表現主義和抽象畫風聞名，作品風格鮮明且充滿視覺衝擊力。

10. Mark Bradford ｜馬克・布拉德福德

英國，以抽象藝術和城市地圖為靈感的作品聞名，風格詳細且充滿視覺深度。

11. Anselm Kiefer ｜安瑟姆・基弗

德國，以新表現主義和歷史題材聞名，作品風格鮮明且充滿視覺衝擊力。

12. Robert Rauschenberg ｜羅伯特・勞森伯格

美國，以新達達主義和混合媒材作品聞名，風格詳細且充滿視覺深度。

13. René Goscinny ｜雷內・戈西尼

法國，以《阿斯特裡克斯》系列聞名，風格幽默且充滿冒險精神。

14. Juanjo Guarnido ｜胡安霍・瓜爾尼多

西班牙，《布拉克薩德》的主要插畫家，以其細膩且生動的畫風聞名。

15. Philippe Druillet ｜菲力浦・德里耶

法國，以科幻漫畫著稱，風格奇幻且充滿細節。

16. Kerry James Marshall 凱瑞・詹姆斯・馬歇爾

英國，以具象繪畫和非裔美國人歷史為題材的作品聞名，風格詳細且充滿視覺衝擊力。

17. Albert Uderzo ｜ 阿爾貝・烏德佐

　法國，《阿斯特裡克斯》的聯合創作者，以其色彩繽紛且細緻的畫風聞名。.

18. Jean-Claude Mézières
　　尚 - 克勞德・梅濟耶

　法國，以《瓦雷裡安與洛琳》系列聞名，風格科幻且富有想像力。

19. Francesco Francavilla
　　法蘭西斯科・弗蘭卡維拉

　義大利，以大膽的線條、憂鬱和黑暗風格著稱，作品充滿視覺衝擊力。

20. Victoria Frances ｜ 維多利亞・法蘭西斯

　西班牙，以浪漫且黑暗的風格聞名，作品充滿情感和視覺深度。

20 位日本插畫家和漫畫家

01. Hirohiko Araki ｜荒木飛呂彥

日本，以《JOJO 的奇妙冒險》系列聞名，風格大膽且具戲劇性。

02. Osamu Tezuka ｜手塚治蟲

日本，被稱為「漫畫之神」，以開創性的漫畫和動畫作品聞名。

03. Fujiko F. Fujio ｜藤子・F・不二雄

日本，以《哆啦 A 夢》系列聞名，風格簡單且充滿童趣。

04. Masakazu Katsura ｜桂正和

日本，以精緻的角色設計和浪漫劇情聞名，作品風格細緻且富有情感。

05. Katsuhiro Otomo │ 大友克洋

│ 日本，以《阿基拉》聞名，風格科幻且充滿細節。

06. Takehiko Inoue │ 井上雄彥

│ 日本，以《灌籃高手》和《浪客劍心》聞名，作品風格細
│ 膩且充滿動感。

07. Makoto Shinkai │ 新海誠

│ 日本，以動畫電影《你的名字》聞名，風格優美且充滿幻想。

08. Taiyo Matsumoto │ 松本大洋

│ 日本，以《鐵兵》等作品聞名，風格獨特且充滿視覺張力。

09. Akira Toriyama ｜ 鳥山明

｜ 日本，以《七龍珠》系列聞名，風格鮮明且富有動感。

10. Gosho Aoyama ｜ 青山剛昌

｜ 日本，以《名偵探柯南》系列聞名，風格細膩且充滿懸疑。

11. Naohisa Inoue ｜ 井上直久

｜ 日本，以幻想風景畫著稱，作品充滿夢幻和神秘色彩。

12. Yoshitomo Nara ｜ 奈良美智

｜ 日本，以簡單而具有情感的童年主題聞名，風格獨特且充滿童真。

13. Akihiko Yoshida ｜吉田明彦

日本，以精緻的角色設計和細膩的畫風著稱，風格鮮明且充滿視覺吸引力。

14. Kentaro Miura ｜三浦建太郎

日本，以《劍風傳奇》系列聞名，作品風格黑暗且富有動感。

15. Onda Naoyuki ｜恩田尚之

日本，風格細膩且充滿張力。

16. Katsuya Terada ｜寺田克也

日本，以插畫和漫畫聞名，風格鮮明且充滿視覺衝擊力。

17. Rumiko Takahashi｜高橋留美子

日本，以《福星小子》和《犬夜叉》等作品聞名，風格幽默且富有情感。

18. NisiOisiN｜西尾維新

日本，以文字和插畫結合作品聞名，風格獨特充滿創意。

09. Mitsumasa Anno｜安野光雅

日本，以兒童繪本和精細畫風著稱，風格細膩充滿童趣。

20. Keiichi Tanaami｜田名網敬一

日本，以多元藝術風格和鮮豔色彩聞名，作品風格大膽且富有創意。

除了可以在提示詞中加入漫畫家及其作品名稱外，也可以利用 Midjourney 內建的功能，在提示詞前輸入參照圖的網址，接下來系統便會根據參照圖的風格結合提示詞生成全新的影像。

卡通動畫風格在 Midjourney 是很重要的一種風格元素，除了在現行的 V6 版本可以有不錯的效果外，Midjourney 中還有個特殊的版本稱為「Niji Model」，Niji 是 Midjourney 和另一家由麻省理工的新創團隊 Spellbrush 共同合作開發，旨在製作動漫和插圖風格，其模型具備更多的動漫、動漫風格和動漫美學知識。它非常適合動態和動作鏡頭以及以角色為中心的構圖。而我們正常使用的 V6.0 版本，仍舊可以針對特定的動漫作品及創作者進行風格定義。

以下我們將列舉各種不同的動漫風格，作為讀者日後生成影像時的參考。在示例圖我們也特別將 V6 版本與 niji 版本進行對比，讀者可視需求選擇適合的版本。

> 提示詞：**animation or carton in the style of** {{ 動畫公司或作品名稱 }}

01. Pixar 皮克斯

以高品質的電腦動畫和深刻的敘事聞名，作品如《玩具總動員》和《海底總動員》。

animation or carton in the style of Pixar **--V 6.0** --style raw　　animation or carton in the style of Pixar **--niji 6** --style raw

02. Disney 迪士尼

以經典的 2D 動畫和現代的 3D 動畫聞名，作品如《獅子王》和《冰雪奇緣》。

03. DreamWorks 夢工廠

以幽默和冒險為主題的動畫電影聞名，作品如《史瑞克》和《馬達加斯加》。

04. Warner Bros. 華納兄弟
| 以經典的卡通角色和幽默風格聞名，作品如《樂一通》和《DC 動畫電影宇宙》。

05. 漢納 - 巴伯拉 Hanna-Barbera
| 以多產的電視動畫節目聞名，作品如《摩登原始人》和《史酷比》。

06. Blue Sky Studios 藍天工作室

以創新的 3D 動畫和幽默聞名，作品如《冰河世紀》和《里約大冒險》。

07. Illumination Entertainment 照明娛樂

以簡單且吸引人的動畫風格聞名，作品如《神偷奶爸》和《寵物當家》。

08. Laika 萊卡

以精細的定格動畫和黑暗童話風格聞名，作品如《鬼媽媽》和《久保與二弦琴》。

09. Cartoon Network 卡通頻道

以創新且多樣的動畫節目聞名，作品如《飛天小女警》和《探險活寶》。

10. Nickelodeon 尼克頻道

以多樣且幽默的動畫節目聞名,作品如《海綿寶寶》和《忍者龜》。

11. The Simpsons 辛普森家庭

以幽默和社會評論聞名,動畫風格簡單且鮮明。

12. Rick and Morty 瑞克和莫蒂

以科幻和黑色幽默聞名，動畫風格大膽且充滿創意。

13. Adventure Time 探險活寶

以奇幻和幽默風格聞名，動畫風格獨特且充滿想像力。

14. Family Guy 蓋酷家庭

以成人幽默和諷刺社會現象聞名，動畫風格簡單且有力。

15. South Park 南方公園

以黑色幽默和尖銳的社會批判聞名，動畫風格粗糙且直接。

16. Gravity Falls 重力泉

以奇幻和神秘題材聞名,動畫風格獨特且有深度。

17. Steven Universe 宇宙小子

以多元文化和深刻的情感描寫聞名,動畫風格鮮明且色彩豐富。

18. The Powerpuff Girls 飛天小女警

以簡單且富有衝擊力的動畫風格聞名，作品充滿了動作和幽默。

19. Bojack Horseman 馬男波傑克

以黑色幽默和深刻的角色描寫聞名，動畫風格獨特且具有成年觀眾的吸引力。

20. 少年悍將 GOTeen Titans Go!
以幽默和快節奏的敘事風格聞名，動畫風格鮮明且有趣。

這些風格涵蓋了美國動畫的廣泛範疇，展示了從經典 2D 到現代 3D 動畫的多樣性和創造力。

20 位日本動畫風格

01. Hayao Miyazaki's Spirited Away
千與千尋──宮崎駿

02. Mamoru Hosoda's Wolf Children
狼的孩子雨和雪──細田守

03. Makoto Shinkai's Your Name
你的名字──新海誠

04. Satoshi Kon's Paprika
   紅辣椒──今敏

05. Isao Takahata's Grave of the Fireflies
   螢火蟲之墓──高畑勳

06. Katsuhiro Otomo's Akira
    阿基拉──大友克洋

07. Hideaki Anno's Neon Genesis Evangelion
    新世紀福音戰士──庵野秀明

08. Masashi Kishimoto's Naruto
火影忍者——岸本齊史

09. Eiichiro Oda's One Piece
航海王——尾田榮一郎

10. Naoko Takeuchi's Sailor Moon
    美少女戰士——武內直子

11. Yoshihiro Togashi's Hunter x Hunter
    獵人 × 獵人——富堅義博

12. Akira Toriyama's Dragon Ball
七龍珠──鳥山明

13. Rumiko Takahashi's Inuyasha
犬夜叉──高橋留美子

191

14. Clamp's Cardcaptor Sakura
庫洛魔法使──Clamp

15. Masamune Shirow's Ghost in the Shell
攻殼機動隊──士郎正宗

16. Tite Kubo's Bleach
死神──久保帶人

17. Hajime Isayama's Attack on Titan
進擊的巨人──諫山創

18. Yasuhiro Nightow's Trigun
槍神──內藤泰弘

19. Tsugumi Ohba & Takeshi Obata's Death Note
死亡筆記──大場鶇 & 小畑健

20. Ken Akamatsu's Love Hina
　　純情房東俏房客──赤松健

這些作品展示了日本動畫在全球的影響力和多樣性，涵蓋了不同的主題和風格。

歐洲知名動畫作品

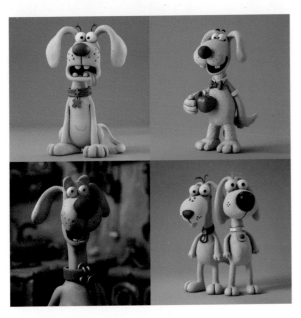

01. Nick Park's Wallace & Gromit
　　超級無敵掌門狗──尼克・派克

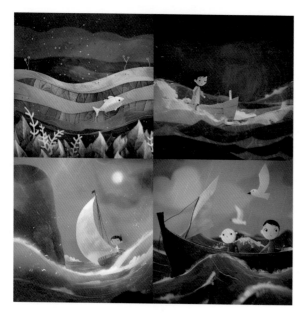

02. Tomm Moore's Song of the Sea
海洋之歌──湯姆・摩爾

03. Sylvain Chomet's The Triplets of Belleville
貝爾維爾三重奏──西維恩・肖梅

04. Marjane Satrapi's Persepolis
   佩爾塞波利斯──瑪嘉・莎塔碧

05. Michel Ocelot's Kirikou and the Sorceress
   奇力庫與巫師──米歇爾・奧斯洛

06. Jean-François Laguionie's The Painting
畫框裡的女人——讓 - 弗朗索瓦·拉格尼

07. Richard Starzak & Mark Burton's Shaun the Sheep Movie
小羊肖恩——大衛·斯帕紮克和馬克·伯頓

08. Claude Barras' My Life as a Zucchini
　　我的人生如西葫蘆──克勞德‧巴拉斯

09. Jérémy Clapin's I Lost My Body
　　我失去了身體──傑瑞米‧克拉平

10. Stefan Fjeldmark's Terkel in Trouble
特克爾的麻煩──斯蒂芬‧菲爾德馬克

11. Bruno Coulais & Alexandre Desplat's The Secret of Kells
凱爾經的秘密──布魯諾‧庫萊斯和亞歷山大‧德斯普拉

12. Marc James Roels & Emma De Swaef's This Magnificent Cake!
這片壯麗的蛋糕——馬克・詹姆斯・羅爾斯和艾瑪・德・斯瓦夫

13. Alberto Vázquez & Pedro Rivero's Birdboy: The Forgotten Children
鳥孩：被遺忘的孩子們——阿爾貝托・瓦斯奎茲和佩德羅・裡維羅

14. Michaël Dudok de Wit's The Red Turtle
紅龜──米迦勒・杜多克・德・維特

15. Alê Abreu's Boy and the World
男孩與世界──阿雷・阿布瑞烏

16. Rémi Chayé's Long Way North
    北方遠征──雷米‧沙耶

17. Anca Damian's Crulic: The Path to Beyond
    克魯利克：前往彼岸之路──安卡‧達米安

18. Igor Kovalyov's Milch
米爾赫──伊戈爾・科瓦廖夫

19. Gisaburo Sugii's The Life of Guskou Budori
小布多的生活──杉井儀三郎

20. Bjørn-Erik Hanssen's Elias: The Little Rescue Boat
　　小救生艇伊萊亞斯──比約恩 - 埃裡克‧漢森

這些作品展示了歐洲動畫的多樣性和創新性，涵蓋了不同的風格和故事題材。

07 電影藝術風格　　提示詞：A movie scene by {{ 電影導演名 }}

電影風格在 2D 平面影像陳述上是一種比較模糊的概念，畢竟電影是一連續性的畫面，但也不同於攝影畫面，通常電影風格的光線表現較平面攝影更加複雜，鏡頭的穿透感也不同，更適合敘事型的影像。在 Midjourney 中設定所謂電影風格，不妨就視為錄影畫面的截圖，我們也模擬了 20 位知名、風格強烈的電影導演，供生成時作為風格參考。

01. Alfred Hitchcock　│希區考克
英國，擅長驚悚與懸疑，風格黑暗且具有強烈的視覺張力。

02. Francis Ford Coppola
　　法蘭西斯‧柯波拉
美國，以史詩電影和家庭劇聞名，作品如《教父》系列。

03. Stanley Kubrick ｜史丹利‧庫柏力克

英國／美國，以視覺和敘事技巧著稱，作品如《2001 太空漫遊》。

04. Martin Scorsese ｜馬丁‧史柯西斯

美國，以犯罪電影和心理劇聞名，作品如《出租車司機》。

05. Quentin Tarantino ｜昆汀‧塔倫提諾

美國，以獨特的對話風格和暴力美學著稱，作品如《低俗小說》。

06. Steven Spielberg ｜史蒂芬‧史匹柏

美國，以冒險和科幻電影聞名，作品如《侏羅紀公園》。

07. Ridley Scott ｜雷利・史考特

英國，以科幻和歷史史詩聞名，作品如《異形》和《角鬥士》。

08. Christopher Nolan ｜克里斯多福・諾蘭

英國，以複雜的敘事結構和視覺效果聞名，作品如《黑暗騎士》三部曲。

09. Tim Burton ｜提姆・波頓

美國，以哥特風格和奇幻題材聞名，作品如《剪刀手愛德華》。

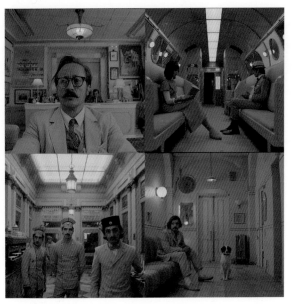

10. Wes Anderson ｜魏斯・安德森

美國，以對稱構圖和獨特的色彩搭配聞名，作品如《布達佩斯大飯店》。

11. David Lynch ｜大衛・林區

美國，以超現實和心理驚悚聞名，作品如《雙峰》。

12. Pedro Almodóvar ｜佩卓・阿莫多瓦

西班牙，以情感強烈的戲劇和鮮豔的色彩聞名，作品如《對她說》。

13. Hayao Miyazaki ｜宮崎駿

日本，以動畫電影聞名，作品如《千與千尋》。

14. Akira Kurosawa ｜黑澤明

日本，以史詩電影和武士題材聞名，作品如《七武士》。

15. Bong Joon-ho｜奉俊昊

| 韓國，以社會議題和黑色幽默聞名，作品如《寄生上流》。

16. Guillermo del Toro｜吉勒摩・戴托羅

| 墨西哥，以奇幻和恐怖題材聞名，作品如《潘神的迷宮》。

17. Jean-Luc Godard｜尚盧・高達

| 法國，以法國新浪潮電影聞名，作品如《筋疲力盡》。

18. Federico Fellini｜費里尼

| 義大利，以超現實和夢幻題材聞名，作品如《八部半》。

19. Ingmar Bergman │英格瑪‧柏格曼
│ 瑞典，以心理劇和存在主義題材聞名，作品如《野草莓》。

20. Andrei Tarkovsky │安德烈‧塔可夫斯基
│ 俄羅斯，以長鏡頭和哲學題材聞名，作品如《索拉裡斯》。

這些導演的風格在 Midjourney 的 AI 生成影像中被廣泛應用，用來創造出各種具有電影感的圖像，這些圖像常常帶有強烈的視覺和情感衝擊力。

例如：我們以 Ingmar Bergman 的風格來生成一張海邊失戀的女人。

poignant and introspective scene of a woman experiencing heartbreak by the seaside, rendered in the style of Ingmar Bergman. The image, white, emphasizing the stark contrasts, shadows typical of Bergman, cinematography. The woman stands alone on a desolate beach --ar 16:9

以英格瑪‧伯格曼的風格呈現一個女人在海邊經歷心碎的淒美而內省的場景。影像是白色的，強調鮮明的對比、陰影，這是伯格曼電影攝影典型。女人獨自站在荒涼的海灘上。

介紹了數百種從不同藝術時期、創作媒材、標的物等各種不同的風格表述後，我們還是要不斷地強調，在目前 AI 平台工具上，**這些藝術家風格的表示，與藝術家本人創作是無任何直接關係，至多可以將這些人名視為某種形式的簡約代名詞**，而我們可以這些名詞作為一個易懂易記的參照符元。

根據風格生成作品，除了媒材、表現手法外，甚麼樣的主題適合搭配甚麼樣的風格，可以得出最完美適切的效果，更是一個學習重點。AI 平台及現有的模型應用不是萬能的，因此在提示詞的主題創作與風格搭配，我們還是建議先追求合理性，最後再嘗試各種不合理及打破常規的變化組合。

在這一個單元，我們先羅列了 50 種不同的插畫風格，但不賦予任何主題，讓 AI 生成最適切的影像定義。然後我們根據這些 AI 生成的風格定義，分別給與不同的主題提示詞，並附上中文翻譯，有助於快速的理解風格與主題的適配性。

> **AI 生成：插畫風格影像 + 需生成的主題提示詞**

> **提示詞：Illustration by {{ 插畫家名 }}**

01. Jen Bartel（珍・巴特爾）

美國，鮮豔色彩，強烈對比，專注於女性角色和女性主義題材。
An illustration in vibrant colors is in the style of Jen Bartel. A strong woman standing in a bustling city street during sunset, wearing a colorful dress --ar 2:3

02. Andy Rementer（安迪‧雷門特爾）

美國，色彩繽紛，異想天開的角色設計，簡約而有趣的圖形風格。
An illustration in colorful and whimsical style is in the style of Andy Rementer. A young girl with her pet dog, sitting in a lively park on a sunny day --ar 2:3

03. Gennady Spirin（根納季‧斯皮林）

俄羅斯，精細的水彩插畫，富有細節的童話和兒童書籍插畫。
A watercolor illustration is in the style of Gennady Spirin. A fairy princess in an enchanted forest during dawn, surrounded by magical creatures --ar 3:2

04. Abbey Lossing（艾比・洛辛）

美國，編輯和雜誌插畫，清晰的線條，柔和的色彩，現代和時尚風格。
An illustration in modern and stylish style is in the style of Abbey Lossing. A businesswoman working in a trendy cafe during morning hours --ar 2:3

05. Albert Dubout（阿爾伯特・杜布）

法國，幽默和戲劇性的插畫，豐富的角色表情和動作，經常帶有諷刺意味。
An illustration with humor and dramatic style is in the style of Albert Dubout. A clumsy chef in a busy kitchen during lunchtime, with exaggerated expressions and chaos --ar 3:2

06. Bob Kane（鮑勃‧凱恩）

美國，蝙蝠俠共同創作者，經典超級英雄漫畫藝術，強烈的陰影和戲劇性動作。
A comic illustration is in the style of Bob Kane. Batman fighting villains in a dark alley at midnight --ar 2:3

07. Shirley Hughes（雪莉‧休斯）

英國，溫暖和細膩的兒童書籍插畫，家庭和日常生活場景，柔和的色彩和細節。
An illustration in warm and detailed style is in the style of Shirley Hughes. A mother and children playing in their cozy living room on a rainy afternoon --ar 3:2

08. Humberto Ramos（翰柏托·拉莫斯）

墨西哥，動感和戲劇性的漫畫書和圖畫小說，強烈的線條和動態構圖。
A dynamic comic illustration is in the style of Humberto Ramos. A superhero in action, leaping between buildings in a vibrant cityscape at noon --ar 2:3

09. Adrienne Segur（阿德里安·塞古）

法國，夢幻且具藝術性的兒童書籍插畫，細緻的細節和優雅的構圖。
An illustration in dreamy and artistic style is in the style of Adrienne Segur. A young girl reading a book in a magical library during evening --ar 3:2

10. James Avati（詹姆斯・阿瓦提）

美國，寫實和戲劇性的書籍封面插畫，強調人物情感和場景的戲劇性。
An illustration in realistic and dramatic style is in the style of James Avati. A man holding a woman in his arms in a passionate embrace, set in a lush garden at sunset --ar 2:3

11. Gustav-Adolf Mossa（古斯塔夫 - 阿道夫・摩薩）

法國，超現實和詳細的插畫，夢幻且略帶不安的場景，富有象徵意義。
An illustration in surreal and detailed style is in the style of Gustav-Adolf Mossa. A mysterious woman in a dreamlike landscape during twilight, surrounded by symbolic elements --ar 3:2

12. Gerald Brom（杰拉德‧布羅姆）

美國，黑暗奇幻風格，細緻且具有戲劇性的場景，經常帶有哥特和蒸汽龐克元素。

An illustration in dark fantasy style is in the style of Gerald Brom. A dark knight standing in a haunted castle during a stormy night, holding a glowing sword --ar 2:3

13. James Thurber（詹姆斯‧瑟伯）

美國，漫畫家，幽默作家，簡單而充滿表情的線條漫畫，幽默且機智的內容。

An illustration in humorous and witty style is in the style of James Thurber. A businessman slipping on a banana peel in an office corridor during lunch break, with exaggerated expressions --ar 3:2

14. Greg Tocchini（格雷格‧托奇尼）

巴西，動感和細緻的漫畫藝術，流動的線條和強烈的動態感。

An illustration in dynamic comic style is in the style of Greg Tocchini. A superheroine soaring through the sky above a futuristic city at sunrise, with fluid lines and dynamic composition --ar 2:3

15. Julie Paschkis（朱莉‧帕斯奇斯）

美國，豐富圖案和色彩的兒童書籍插畫，靈感來自民間故事和傳說。

An illustration in colorful and patterned style is in the style of Julie Paschkis. A young boy discovering a hidden treasure in a mystical forest during twilight, surrounded by mythical creatures --ar 3:2

16. Gregoire Guillemin（格雷戈爾·吉爾曼）

法國，流行藝術，日常圖像的再創作，明亮的色彩和平滑的線條。
An illustration in pop art style is in the style of Gregoire Guillemin. A woman sipping coffee in a retro diner during the 1960s, with bright colors and smooth lines --ar 2:3

17. Greg Hildebrandt（格雷格·希爾德布蘭特）

美國，奇幻和科幻插畫，細緻的細節和豐富的色彩，著名於《魔戒》插畫。
An illustration in fantasy and sci-fi style is in the style of Greg Hildebrandt. A group of adventurers exploring an alien planet during a sunny day, with vibrant alien flora --ar 3:2

18. J.P. Targete（J.P. 塔吉特）

美國，科幻和奇幻插畫，細緻的細節和戲劇性的光影效果。
An illustration in sci-fi and fantasy style is in the style of J.P. Targete. A wizard casting a spell in a futuristic city during dusk, with dramatic lighting and intricate details --ar 2:3

19. H.R. Giger（H.R. 吉格爾）

瑞士，超現實主義和恐怖風格，生物機械學的藝術，暗色調和怪異的形態。
An illustration in surreal and biomechanical style is in the style of H.R. Giger. A cyborg exploring an abandoned spaceship in deep space during a dark night, surrounded by eerie structures --ar 3:2

20. Jim Lee（吉姆・李）

韓裔美國，超級英雄漫畫藝術，強烈的動態構圖和戲劇性陰影。
An illustration in superhero comic style is in the style of Jim Lee. A superhero leaping from a skyscraper in a bustling city at night, with dynamic composition and dramatic shadows --ar 2:3

21. James Paick（詹姆斯・帕克）

美國，概念藝術和數字繪畫，科幻和奇幻主題，豐富的色彩和細節。
A digital painting in concept art style is in the style of James Paick. A team of explorers navigating through a dense jungle on an alien planet during dawn, with vibrant colors and intricate details --ar 3:2

22. Kerem Beyit（克雷姆‧貝伊特）

土耳其，奇幻和科幻插畫，細緻的細節和豐富的色彩，經常描繪英雄和奇幻生物。
An illustration in fantasy and sci-fi style is in the style of Kerem Beyit. A mighty hero battling a giant robot in a futuristic arena during noon, with detailed and vibrant colors --ar 2:3

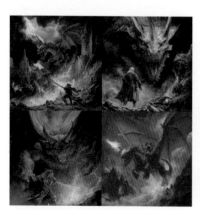

23. Jeff Easley（傑夫‧伊斯利）

美國，奇幻插畫，D&D（龍與地下城）風格，細緻的細節和豐富的色彩。
An illustration in fantasy style is in the style of Jeff Easley. A dragon battling a knight on a misty mountain peak during dawn, with detailed and vibrant colors --ar 3:2

24. Kinuko Y. Craft（金子‧Y‧克拉夫特）

美國，童話和奇幻插畫，精美的細節和華麗的構圖，靈感來自經典文學和神話故事。

An illustration in fairy tale style is in the style of Kinuko Y. Craft. A princess in an elaborate gown standing in a magical garden during twilight, with intricate details and vibrant colors --ar 2:3

25. Jim Burns（吉姆‧伯恩斯）

英國，科幻插畫，細緻的細節和宏大的場景，經常描繪未來科技和宇宙景觀。

An illustration in sci-fi style is in the style of Jim Burns. A spaceship landing on a distant planet with alien flora, during twilight, with intricate details --ar 3:2

26. Luis Royo（路易斯・羅約）

西班牙，黑暗奇幻和超現實主義插畫，細緻的細節和戲劇性的光影效果。
An illustration in dark fantasy style is in the style of Luis Royo. A warrior queen standing on a battlefield at dusk, with dramatic lighting and intricate armor --ar 2:3

27. John Berkey（約翰・貝基）

美國，科幻插畫，宏大的宇宙場景和未來科技，富有細節和戲劇性。
An illustration in sci-fi style is in the style of John Berkey. A fleet of spaceships navigating through an asteroid field during a cosmic storm, with detailed and dramatic composition --ar 3:2

28. Magali Villeneuve（瑪加莉・維倫紐夫）

法國，奇幻和科幻插畫，細緻的細節和豐富的色彩，經常描繪英雄和奇幻生物。

An illustration in fantasy and sci-fi style is in the style of Magali Villeneuve. A hero battling an alien creature in an enchanted forest during sunset, with detailed and vibrant colors --ar 2:3

29. John Blanche（約翰・布蘭奇）

英國，黑暗奇幻和科幻插畫，細緻的細節和暗色調，經常帶有蒸汽龐克元素。

An illustration in dark fantasy and steampunk style is in the style of John Blanche. A steampunk warrior standing in a dystopian cityscape at dusk, with intricate details and dark tones --ar 3:2

30. Mlanie Delon（梅拉妮・德隆）

法國，奇幻與寫實插畫，詳細且情感豐富的人物與場景描繪。

An illustration in fantasy realism style is in the style of Mélanie Delon. A sorrowful princess sitting by a serene lake at dawn, with detailed expressions and rich colors --ar 2:3

31. John Harris（約翰・哈里斯）

英國，宏大的科幻插畫，壯麗的宇宙景觀和未來建築，柔和的色彩和光影效果。

An illustration in grand sci-fi style is in the style of John Harris. A futuristic city with towering buildings under a starry sky during twilight, with soft colors and light effects --ar 3:2

32. Nene Thomas（妮妮‧托馬斯）

美國，奇幻插畫，詳細且華麗的人物和環境描繪。

An illustration in detailed and ornate fantasy style is in the style of Nene Thomas. A fairy queen standing on a cliff overlooking a magical valley at sunrise, with intricate details and rich colors --ar 2:3

33. John Howe（約翰‧豪）

加拿大，奇幻插畫，特別是《魔戒》系列，細緻的細節和戲劇性的場景。

An illustration in fantasy style is in the style of John Howe. A band of adventurers trekking through a mystical forest at dawn, with detailed and dramatic scenery --ar 3:2

34. Brian Froud（布萊恩 · 弗勞德）

英國，奇幻插畫，特別是妖精和神話生物，豐富的細節和質感。

An illustration in whimsical fantasy style is in the style of Brian Froud. A group of fairies dancing around a magical tree in an enchanted forest during twilight, with detailed textures and mythical creatures --ar 3:2

35. John Schoenherr（約翰 · 舍恩赫爾）

美國，科幻和奇幻插畫，細緻的細節和豐富的色彩，經常描繪未來世界和奇幻生物。

An illustration in sci-fi and fantasy style is in the style of John Schoenherr. An alien creature exploring a vibrant alien jungle at sunset, with rich colors and intricate details --ar 3:2

36. Trina Schart Hyman（特里娜 · 夏特 · 海曼）

美國，兒童書籍插畫，溫暖且詳細的人物和場景描繪。
An illustration in warm and detailed children's book style is in the style of Trina Schart Hyman. A family gathered around a fireplace in a cozy living room during winter evening, with warm colors and intricate details --ar 3:2

37. Mat Collishaw（馬特 · 科利肖）

英國，當代藝術，富有創意和挑戰性的作品，探索社會和文化主題。
An illustration in contemporary art style is in the style of Mat Collishaw. A provocative piece depicting a clash of cultures in a modern city during the day, with bold colors and strong symbolism --ar 3:2

38. TonyDiTerlizzi（托尼‧迪特里茲）

美國，奇幻與兒童書籍插畫，知名於《蜘蛛與我》系列，豐富的細節和想像力。

An illustration in imaginative children's book style is in the style of Tony DiTerlizzi. A young boy discovering a hidden world in his backyard during a sunny afternoon, with rich details and fantastical elements --ar 2:3

39. Karel Appel（卡雷爾‧阿佩爾）

荷蘭，表現主義和抽象藝術，大膽的色彩和粗獷的筆觸，充滿情感的作品。

An abstract painting in expressionist style is in the style of Karel Appel. A dynamic explosion of colors and shapes depicting a chaotic cityscape, with bold strokes and emotional intensity --ar 3:2

40. Larry Elmore（拉里 Kinuko Y. Craft 埃爾莫爾）

美國，奇幻插畫，D&D（龍與地下城）風格，細緻的細節和動感的構圖。

An illustration in fantasy style is in the style of Larry Elmore. A group of adventurers fighting a dragon in a dark cave, with dynamic composition and detailed characters --ar 3:2

41. Maciej Kuciara（馬切伊・庫希亞拉）

波蘭，概念藝術和數字繪畫，科幻和奇幻主題，豐富的色彩和細節。

A digital painting in concept art style is in the style of Maciej Kuciara. A futuristic cityscape with flying cars during a bright day, with vibrant colors and detailed architecture --ar 3:2

42. Max Ernst（馬克斯・恩斯特）

德國，超現實主義，大膽的構圖和夢幻的場景，經常帶有神秘和象徵主義元素。
An illustration in surrealist style is in the style of Max Ernst. A dreamlike landscape with mysterious figures during twilight, filled with symbolic elements and bold compositions --ar 3:2

43. Michael Komarck（邁克爾・科馬克）

美國，奇幻與科幻插畫，豐富的細節和戲劇性的構圖。
An illustration in fantasy and sci-fi style is in the style of Michael Komarck. A starship captain commanding a fleet in deep space during a cosmic battle, with dramatic lighting and intricate details --ar 3:2

44. Michael Whelan（邁克爾・惠蘭）

美國，科幻與奇幻插畫，鮮豔的色彩和複雜的細節。
An illustration in sci-fi and fantasy style is in the style of Michael Whelan. An explorer discovering an ancient alien artifact in a lush jungle during a bright day, with vibrant colors and detailed scenery --ar 3:2

45. Mike Azevedo（麥克・阿澤維多）

巴西，動感且色彩豐富的插畫，強烈的光影和豐富的角色設計。
An illustration in dynamic and colorful style is in the style of Mike Azevedo. A brave warrior fighting a giant monster in a mystical forest during dusk, with vibrant colors and dramatic lighting --ar 3:2

46. Moebius（尚・吉羅 / 墨比斯）

法國，奇幻與科幻插畫，流暢的線條和富有想像力的世界。

An illustration in fantastical sci-fi style is in the style of Moebius. A lone traveler exploring a surreal desert landscape under a vibrant sky, with smooth lines and imaginative elements --ar 3:2

47. Frank Frazetta（法蘭克・佛拉薩塔）

美國，奇幻與科幻插畫，強壯的人物和戲劇性的動作場景。

An illustration in muscular and dramatic fantasy style is in the style of Frank Frazetta. A barbarian warrior battling enemies on a rugged battlefield during a storm, with powerful poses and dynamic action --ar 3:2

48. Arthur Rackham（亞瑟・拉克姆）

英國，童話與奇幻插畫，精細的線條和氛圍濃厚的構圖。

An illustration in fairy tale style is in the style of Arthur Rackham. A young girl meeting a talking animal in a dark, mysterious forest during nightfall, with intricate line work and atmospheric composition --ar 2:3

49. Beatrix Potter（比阿特麗克斯・波特）

英國，兒童書籍插畫，以彼得兔系列最為著名，迷人且詳細的作品。

An illustration in charming children's book style is in the style of Beatrix Potter. Peter Rabbit exploring a beautiful garden during a spring morning, with detailed and charming scenery --ar 2:3

50. Kay Nielsen（凱‧尼爾森）

丹麥，童話與奇幻插畫，華麗的裝飾風格，靈感來自斯堪的納維亞的民間故事。

An illustration in decorative fairy tale style is in the style of Kay Nielsen. A scene from a Scandinavian folktale with intricate patterns and elegant characters during twilight, with rich colors and ornate details --ar 3:2

三、構成（Constitute）

除了上述的媒材、風格外，這個章節我們來談畫面的構成。在多數時候，如果不加以構圖的自行定義，AI 會根據主題或風格進行其最佳化構圖，例如風景多採用廣角，城市夜景多採用鳥瞰視角。但除了讓 AI 自行定義外，我們更希望一個畫面是能夠 100% 由自己控制的，因此在這單元我們將針對視角（構圖）、色彩描述及畫面氛圍做更詳盡的定義說明。

▶ 視角（構圖） 🖥 237

善用視角定義可以使影像生成更具創意和表現力。不同的視角和鏡頭角度能夠改變主體的呈現方式，傳達不同的情感和故事。例如，低角度可以讓主體顯得更有威嚴，而高角度則可以表現出脆弱感。透過精確的視角描述，您可以更好地控制影像的構圖和氛圍，達到理想的視覺效果。

以下是常使用的的視角定義：

01. Eye-Level Shot｜眼平視角

鏡頭位於主體的眼睛水平，創造出自然、平衡的視角。
Eye-level shot of a person walking through a busy market.
--ar 3:2
眼平視角，一個人在繁忙的市場中行走。

02. Low Angle｜低角度

鏡頭位於主體的眼睛下方，向上拍攝，使主體看起來更大、更具威嚴。
Low-angle shot of a towering ancient statue. --ar 2:3

03. High Angle ｜高角度

鏡頭於主體上方向下拍攝，使主體看起來較小或更脆弱。
High-angle shot of a child playing in a large park --ar 3:2

04. Bird's Eye View, Top-down View 鳥瞰視角或俯視視角

從高處直接向下拍攝，提供全面的場景視圖。
Bird's-eye view of a city grid at night --ar 3:2

05. Worm's Eye View ｜蟲瞰視角

鏡頭在地面水平向上拍攝，強調高度，使主體看起來宏偉。
Worm's-eye view of tall trees in a dense forest. --ar 3:2

06. Wide Shot ｜廣角鏡

拍攝大範圍場景，建立環境或設定背景。
Wide shot of a mountain range at sunrise. --ar 3:2

07. Close-Up ｜特寫鏡頭

近距離拍攝主體，捕捉細節或表情。
Close-up of a person's eyes showing emotion. --ar 3:2

08. Extreme Close-Up ｜極端特寫鏡頭

集中在非常小的細節或特徵上，創造戲劇性強調。
Extreme close-up of a droplet of water on a leaf. --ar 3:2

09. Over-the-Shoulder Shot ｜肩上視角

從角色背後拍攝，展示他們所看到的內容。
Over-the-shoulder shot of a person viewing a cityscape.
--ar 3:2

10. Side View ｜側面視角

展示對象的側面輪廓。這種視角常用於展示動態或對象的側面特徵。
A side view of a running horse, capturing its motion and muscular structure.. --ar 3:2

11. Panoramic View ｜全景視角

寬幅拍攝，捕捉廣闊的景色。使用全景視角時，最好配合長寬比大於 3:2 參數設定，例如 --ar 2:1
Panoramic view of a coastline during sunset. --ar 2:1

12. Macro Shot ｜微距鏡頭

極近距離拍攝小型主體或細節，常用於自然攝影。
Macro shot of the intricate patterns on a butterfly's wing.
--ar 3:2

13. Fisheye Lens ｜魚眼鏡頭

使用超廣角鏡頭，創造出圓形失真效果，覆蓋極寬視角。
Fisheye lens shot of a person standing in the center of a circular building. --ar 3:2

239

使用這些視角設定，您可以在 Midjourney 中創造出動態、引人入勝和視覺上吸引人的圖像，符合您的創意願景。除了視角外，我們也可利用攝影主題範圍（Shot Sizes）的概念應用在提示詞內，便可以生成不同主題大小的構圖。

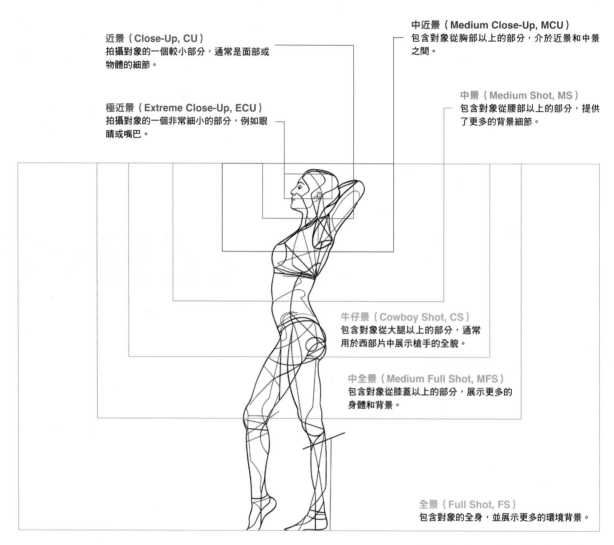

近景（Close-Up, CU）
拍攝對象的一個較小部分，通常是面部或物體的細節。

極近景（Extreme Close-Up, ECU）
拍攝對象的一個非常細小的部分，例如眼睛或嘴巴。

中近景（Medium Close-Up, MCU）
包含對象從胸部以上的部分，介於近景和中景之間。

中景（Medium Shot, MS）
包含對象從腰部以上的部分，提供了更多的背景細節。

牛仔景（Cowboy Shot, CS）
包含對象從大腿以上的部分，通常用於西部片中展示槍手的全貌。

中全景（Medium Full Shot, MFS）
包含對象從膝蓋以上的部分，展示更多的身體和背景。

全景（Full Shot, FS）
包含對象的全身，並展示更多的環境背景。

參考資料：Camera Shot Size Cheat-Sheet

▶ 色彩 🖥 240

在 Midjourney 中使用色彩定義可以讓你更精確地控制生成圖像的外觀。以下是一些基本步驟和技巧來幫助你在 Midjourney 中使用色彩定義：

◆ 使用簡單的色彩描述詞

你可以在提示中使用簡單的色彩描述詞來設定圖像的主色調或精準地替物體上色。例如：紅色（red）、藍色（blue）、綠色（green）、黃色（yellow）、紫色（purple）等，但由於現有 AI 生

成平台對於顏色代碼（如 HEX 或 RGB）的支援並不完整，在顏色的定義上，比較建議用顏色的名稱加入提示詞中，以下提供 120 種常用的色彩名稱以及 HEX 顏色代碼對照。

Reds and Pinks 紅色

圖檔	English	Chinese	HEX Code
	Crimson	緋紅色	#DC143C
	Dark Red	深紅色	#8B0000
	Deep Pink	深粉色	#FF1493
	Firebrick	耐火磚色	#B22222
	Hot Pink	豔粉色	#FF69B4
	Indian Red	印度紅	#CD5C5C
	Light Pink	淺粉色	#FFB6C1
	Medium Violet Red	中紫紅	#C71585
	Misty Rose	霧玫瑰色	#FFE4E1
	Pale Violet Red	淡紫紅	#DB7093
	Pink	粉色	#FFC0CB
	Red	紅色	#FF0000
	Rosy Brown	玫瑰褐	#BC8F8F
	Salmon	鮮肉色	#FA8072
	Tomato	番茄紅	#FF6347

Oranges and Browns 棕色調

圖檔	English	Chinese	HEX Code
	Bisque	桔黃色	#FFE4C4
	Blanched Almond	白杏仁色	#FFEBCD
	Burly Wood	結實的木頭色	#DEB887
	Chocolate	巧克力色	#D2691E
	Coral	珊瑚色	#FF7F50
	Cornsilk	玉米絲色	#FFF8DC
	Dark Orange	深橙色	#FF8C00
	Light Coral	淺珊瑚色	#F08080
	Orange	橙色	#FFA500
	Orange Red	橙紅	#FF4500
	Peach Puff	桃色	#FFDAB9
	Peru	秘魯色	#CD853F
	Sandy Brown	沙褐	#F4A460
	Sienna	赭色	#A0522D
	Tan	棕褐色	#D2B48C

Yellows 黃色調

圖檔	English	Chinese	HEX Code
	Antique White	古董白	#FAEBD7
	Beige	米色	#F5F5DC
	Dark Goldenrod	深金黃	#B8860B
	Gold	金色	#FFD700
	Goldenrod	金棒色	#DAA520
	Khaki	卡其色	#F0E68C
	Lemon Chiffon	檸檬綢色	#FFFACD
	Light Goldenrod Yellow	淺金黃	#FAFAD2
	Light Yellow	淺黃色	#FFFFE0
	Pale Goldenrod	淡金黃	#EEE8AA
	Papaya Whip	番木色	#FFEFD5
	Yellow	黃色	#FFFF00

Greens 綠色調

圖檔	English	Chinese	HEX Code
	Chartreuse	黃綠色	#7FFF00
	Dark Green	深綠色	#006400
	Dark Olive Green	深橄欖綠	#556B2F

圖檔	English	Chinese	HEX Code
	Dark Sea Green	深海綠色	#8FBC8F
	Forest Green	森林綠	#228B22
	Green	綠色	#008000
	Green Yellow	綠黃色	#ADFF2F
	Honeydew	蜜瓜色	#F0FFF0
	Lawn Green	草坪綠	#7CFC00
	Light Green	淺綠色	#90EE90
	Lime	青檸色	#00FF00
	Lime Green	青檸綠	#32CD32
	Medium Aquamarine	中海藍	#66CDAA
	Medium Sea Green	中海綠	#3CB371
	Mint Cream	薄荷奶油	#F5FFFA
	Olive	橄欖色	#808000
	Olive Drab	橄欖褐	#6B8E23
	Pale Green	淡綠色	#98FB98
	Sea Green	海綠	#2E8B57
	Spring Green	春綠色	#00FF7F
	Yellow Green	黃綠色	#9ACD32

Cyans and Blues 藍色調

圖檔	English	Chinese	HEX Code
	Aquamarine	碧綠色	#7FFFD4
	Azure	天藍色	#007FFF
	Cadet Blue	軍校藍	#5F9EA0
	Cornflower Blue	矢車菊藍	#6495ED
	Cyan	青色	#00FFFF
	Dark Blue	深藍色	#00008B
	Dark Cyan	深青色	#008B8B
	Dark Slate Blue	深石板藍	#483D8B
	Dark Turquoise	深綠松石色	#00CED1
	Deep Sky Blue	深天藍色	#00BFFF
	Dodger Blue	道奇藍	#1E90FF
	Light Blue	淺藍色	#ADD8E6
	Light Cyan	淺青色	#E0FFFF
	Light Sea Green	淺海綠	#20B2AA
	Light Sky Blue	淺天藍	#87CEFA
	Light Steel Blue	淺鋼藍	#B0C4DE
	Navy	海軍藍	#000080

圖檔	English	Chinese	HEX Code
	Pale Turquoise	淡綠松石	#AFEEEE
	Powder Blue	粉藍	#B0E0E6
	Royal Blue	皇家藍	#4169E1
	Sky Blue	天藍	#87CEEB
	Slate Blue	石板藍	#6A5ACD
	Slate Gray	石板灰	#708090
	Steel Blue	鋼藍	#4682B4
	Teal	水鴨色	#008080
	Turquoise	綠松石	#40E0D0

Purples 紫色調

圖檔	English	Chinese	HEX Code
	Blue Violet	藍紫色	#8A2BE2
	Dark Magenta	深洋紅	#8B008B
	Dark Orchid	深蘭花紫	#9932CC
	Dark Violet	深紫羅蘭色	#9400D3
	Indigo	靛藍色	#4B0082
	Lavender	薰衣草色	#E6E6FA
	Lavender Blush	淡紫色	#FFF0F5

圖檔	English	Chinese	HEX Code
	Orchid	蘭花紫	#DA70D6
	Plum	梅紅	#DDA0DD
	Purple	紫色	#800080
	Thistle	薊色	#D8BFD8
	Violet	紫羅蘭	#EE82EE

Neutrals 中性色彩

圖檔	English	Chinese	HEX Code
	Alice Blue	愛麗絲藍	#F0F8FF
	Black	黑色	#000000
	Blue	藍色	#0000FF
	Brown	棕色	#A52A2A
	Dim Gray	暗灰色	#696969
	Floral White	花白	#FFFAF0
	Gainsboro	庚斯博羅色	#DCDCDC
	Ghost White	幽靈白	#F8F8FF
	Gray	灰色	#808080
	Ivory	象牙色	#FFFFF0
	Light Gray	淺灰色	#D3D3D3

圖檔	English	Chinese	HEX Code
	Linen	亞麻色	#FAF0E6
	Maroon	栗色	#800000
	Navajo White	納瓦霍白	#FFDEAD
	RosyBrown	粉褐色	#BC8F8F
	Silver	銀色	#C0C0C0
	Snow	雪色	#FFFAFA
	White	白色	#FFFFFF

◆ 指定具體的色調

如果你希望圖像中的某些元素具有特定的色調，也可以具體描述。例如：淺藍色（light blue）、深綠色（dark green）、淡粉色（pale pink）

生成範例｜美麗的日落

A beautiful sunset with red and orange hues
美麗的日落，帶有紅色和橙色的色調

範例

A deep green forest with a light blue river flowing through it
深綠色的森林，淺藍色的河流流經其中。

◆ 結合色彩與物體

描述圖像時，將色彩與具體的物體結合可以讓 Midjourney 更好地理解你的需求。

A red sports car parked under a bright yellow tree
一輛紅色跑車停在一棵亮黃色的樹下。

◆ 使用形容詞增強效果

使用一些形容詞來增強圖像的色彩效果，如：鮮豔的（vibrant）、柔和的（soft）、暗淡的（muted）、強烈的（intense）。

A vibrant red apple on a soft green background
柔和的綠色背景充滿活力的紅色蘋果。

這些方法都可以更好地控制和定義圖像的色彩，使生成的圖像更符合你的期望。最後我們來嘗試一張複雜顏色的設定：緋紅色的長髮女子，庚斯博羅色的上衣，納瓦霍白的長裙，愛麗絲藍的披巾，檸檬綢色的洋傘，耐火磚色的長靴。

In a realistic photography style, a full-body shot of a woman with long crimson hair, wearing a Gainsboro-colored top, a Navajo White long skirt, and an Alice Blue shawl. She is holding a lemon chiffon-colored parasol and wearing firebrick-colored boots. The photo is taken with a Hasselblad camera using a wide-angle lens. The scene is set in a picturesque garden with blooming flowers and lush greenery. --style raw --v 6.0 --ar 2:3

燈光,在傳統攝影中,曝光、光線或是燈光是成像優劣及氛圍營造的關鍵因素。在 AI 影像生成中,燈光風格 Tokens 同樣扮演著至關重要的角色。這些 Tokens 允許使用者指定影像中的光照條件和效果,從而控制影像」的情感和視覺效果。

燈光風格可以模擬不同的光照條件,如日光、黃昏、夜晚、室內燈光等,並根據這些條件調整影像中的光影效果。例如,使用「柔和光」(Soft Light)可以創造出溫暖、柔和的氛圍,適合家庭場景或溫馨的情景;而使用「硬光」(Hard Light)則可以增加影像的對比度和戲劇性,適合動感或強烈情感的場景。

此外,燈光風格 Tokens 還可以模擬特定的光源位置和方向,如背光、側光、頂光等,進一步影響影像的氛圍和視覺效果。例如,背光(Backlight)可以創造出剪影效果,增添神秘和戲劇性;而側光(Side Light)則可以強調對象的輪廓和質感,增強影像的立體感。

以下我們列舉了 40 種不同的場景燈光和燈光效果,從而實現特定的情感表達和視覺效果。在 Midjourney 和其他 AI 影像生成平台中,靈活運用燈光風格,不僅能提升影像的質量和美感,還能幫助創作者更好地傳達其創意和故事。

01. Golden Hour｜黃金時刻
黎明後的第一個小時和黃昏前的最後一個小時
Photo of a serene landscape, Golden Hour

02. Backlit｜逆光
光源來自主體的後方
Photo of a silhouette of a tree, Backlit

03. Diffused Light ｜散射光

柔和均勻的光源
Photo of a woman with soft features, Diffused Light

04. Seasonal Lighting
　　 季節性燈光

反映不同季節特性的光源
Photo of a park in winter, Seasonal Lighting

05. Shadowed ｜陰影光

強烈陰影與亮光對比
Photo of a man with strong features, Shadowed

06. Sunflare ｜太陽耀斑

太陽光透過鏡頭形成的耀斑
Photo of a landscape, Sunflare

07. Rim Lighting ｜邊緣光

燈光照亮主體邊緣
Photo of a woman with rim lighting, Rim Lighting

08. Cross Lighting ｜交叉光

燈光交叉照射
Photo of a man with chiseled features, Cross Lighting

09. High Key Lighting ｜高調光

明亮且陰影最小
Photo of a cheerful child, High Key Lighting

10. Low Key Lighting ｜低調光

深色背景和強烈陰影
Photo of an elderly person, Low Key Lighting

11. Broad Lighting ｜寬光源

寬廣均勻的照明
Photo of a person with broad lighting, Broad Lighting

12. Short Lighting ｜短光源

燈光主要照在臉的一側
Photo of a woman with short lighting, Short Lighting

13. Flashlamp Lighting ｜
閃燈光

由閃光燈提供的快速明亮照明
Photo of a man in action, Flashlamp Lighting

14. Flashtube Lighting ｜
閃光管光

明亮且快速的閃光燈照明
Photo of a woman in studio, Flashtube Lighting

15. Incandescent Lamp ｜ 白熾燈光
暖色的室內白熾燈照明
Photo of a family indoors, Incandescent Lamp

16. Natural Light ｜ 自然光
自然環境中的光源
Photo of a couple outdoors, Natural Light

17. Candlelight ｜ 燭光
溫暖的燭光照明
Photo of a dinner scene with couple, Candlelight

18. Neon Lighting ｜ 霓虹燈光
由霓虹燈提供的亮麗燈光
Photo of a person on a city street, Neon Lighting

19. Spotlight ｜聚光燈

集中照明在特定主體
Photo of a performer on stage, Spotlight

20. Ambient Light ｜環境光

整體環境提供的柔和光源
Photo of a tranquil room with a person, Ambient Light

21. Overhead Light ｜頂光

光源來自頭頂上方
Photo of a person lit from above, Overhead Light

22. Side Lighting ｜側光

光源來自主體的側面
Photo of a person with side lighting, Side Lighting

23. Front Lighting ｜前光

光源直接照在主體的前面
Photo of a person lit from the front, Front Lighting

24. Butterfly Lighting ｜蝴蝶光

在鼻子下方形成蝴蝶狀陰影的照明
Photo of a glamour portrait, Butterfly Lighting

25. Loop Lighting ｜環形光

在鼻子下方形成環狀陰影的照明
Photo of a classic portrait, Loop Lighting

26. Split Lighting ｜分割光

臉部一側在光中，另一側在陰影中的照明
Photo of a person with half face in shadow, Split Lighting

27. Chiaroscuro ｜明暗對比光

強烈的明暗對比照明
Photo of a dramatic scene, Chiaroscuro

28. Window Light ｜窗光

來自窗戶的自然光照明
Photo of a person by a window, Window Light

29. Strobe Lighting ｜速閃光

由速閃燈提供的明亮瞬間照明
Photo of an athlete in action, Strobe Lighting

30. Colored Gels ｜彩色濾光片

使用彩色濾光片創造的鮮豔光效
Photo of a creative portrait, Colored Gels

31. Dappled Light ｜斑駁光

光透過樹葉等形成的斑駁效果
Photo of a forest scene, Dappled Light

32. Moonlight ｜月光

柔和的藍色月光照明
Photo of a night scene with person, Moonlight

33. Blue Hour ｜藍色時刻

日落後和日出前的藍色色溫
Photo of a cityscape at dusk, Blue Hour

34. Prismatic Lighting ｜稜鏡光

創造稜鏡效果的光照明
Photo of a surreal scene, Prismatic Lighting

35. Holographic Lighting ｜全息燈光

產生全息效果的燈光
Photo of a futuristic scene, Holographic Lighting

36. Sunbeams ｜光束

透過樹林等形成的光束
Photo of a forest clearing, Sunbeams

37. Direct Sunlight ｜直射陽光

直接來自太陽的強光
Photo of a surfer, Direct Sunlight

38. Foggy/Misty Lighting ｜霧光

在霧中產生的柔和光照明
Photo of a bridge in fog, Foggy/Misty Lighting

39. Rembrandt Lighting │ 雷姆布蘭特光

│ 創造三角形光斑的側光照明
│ Photo of a person with Rembrandt Lighting

40. Overcast │ 陰天光

│ 陰天條件下的柔和光照明
│ Photo of a beach scene, Overcast

▶ 表情 🖥 257

在 AI 影像生成特別是在創造人物時，表情（Emotion Style）Tokens 是控制影像情感和氛圍非常重要的一環。這些 Tokens 允許使用者指定影像中人物或角色的情感表達，從而增強影像的情感深度和視覺吸引力。

表情 Tokens 可以模擬各種情感狀態，如喜悅、悲傷、憤怒、驚訝、恐懼等。通過選擇和應用這些表情 Tokens，使用者能夠創建出具有豐富情感層次的影像。例如，使用「喜悅」（Joyful）Token 可以讓影像中的人物展現出開心的笑容，營造出愉快、積極的氛圍；而使用「悲傷」（Sad）Token 則可以讓人物顯得憂鬱、傷感，適合表達哀傷或沉重的主題。

此外，表情還可以結合其他風格使用，以達到更複雜的情感表達。例如，結合「驚訝」（Surprised）表情 Token 和「夢幻」（Dreamy）氛圍 Token，可以創造出一幅人物在奇幻場景中驚訝探索的影像，增強故事性和視覺效果。

01. Determined ｜堅定
A determined little girl, portrait in the style of Norman Rockwell

02. Happy ｜快樂
A happy little girl, photograph by Steve McCurry

03. Sleepy ｜困倦
A sleepy little girl, painting in the style of Mary Cassatt

04. Angry ｜生氣
An angry little girl, photograph by Richard Avedon

05. Shy ｜害羞
A shy little girl, painting in the style of Pierre-Auguste Renoir

06. Sad ｜悲傷
A sad little girl, photograph by Sebastião Salgado

07. Excited ｜興奮
An excited little girl, painting in the style of Claude Monet

08. Surprised ｜驚訝
A surprised little girl, photograph by Henri Cartier-Bresson

09. Nervous ｜緊張
A nervous little girl, painting in the style of Edvard Munch

10. Confident ｜自信
A confident little girl, photograph by Annie Leibovitz

11. Thoughtful ｜深思
A thoughtful little girl, painting in the style of Leonardo da Vinci

12. Amused ｜覺得好笑
An amused little girl, photograph by Martin Parr

13. Disgusted ｜厭惡
A disgusted little girl, painting in the style of Vincent van Gogh

14. Scared ｜害怕
A scared little girl, photograph by Dorothea Lange

15. Pensive ｜沉思
A pensive little girl, painting in the style of Johannes Vermeer

16. Content ｜滿足
A content little girl, photograph by Dorothea Lange

17. Bored ｜無聊
A bored little girl, painting in the style of Gustav Klimt

18. Embarrassed ｜尷尬
An embarrassed little girl, photograph by Diane Arbus

19. Curious ｜好奇
A curious little girl, painting in the style of Albrecht Dürer

20. Hopeful ｜充滿希望
A hopeful little girl, photograph by Ansel Adams

21. Frustrated ｜挫敗
A frustrated little girl, painting in the style of Egon Schiele

22. Jealous ｜嫉妒
A jealous little girl, photograph by Cindy Sherman

23. Proud ｜自豪
A proud little girl, painting in the style of Diego Rivera

24. Relieved ｜如釋重負
A relieved little girl, photograph by Irving Penn（圖：EMO024）

25. Anxious ｜焦慮
An anxious little girl, painting in the style of Francisco Goya

26. Enthusiastic ｜熱情
An enthusiastic little girl, photograph by Bruce Weber

27. Mischievous ｜頑皮
A mischievous little girl, painting in the style of Paul Klee

28. Shocked ｜震驚
A shocked little girl, photograph by Bill Brandt

29. Melancholic ｜憂鬱
A melancholic little girl, painting in the style of Edward Hopper

30. Contemplative ｜沉思的
A contemplative little girl, photograph by Walker Evans

▶ 場景 🖥 265

場景（Locations）Tokens 是用來設定影像背景和場景的重要元素。使用者指定影像中的具體場景，從而增強影像的整體情感表達和視覺吸引力，使得角色的表情和情感表達更加具體和具體。不同的場景可以傳達不同的情感氛圍，並對影像的情感和故事產生深遠的影響。例如，在一個溫馨的家庭場景中，角色的微笑會傳達出幸福和安全感，而在一個陰暗的森林場景中，角色的恐懼表情則會加強緊張和恐懼的氣氛。

應用場景 Tokens，可以幫助使用者更精確地控制影像的情感氛圍。例如，選擇「城市街道」作為場景，可以搭配角色的興奮表情，展示都市生活的活力和快節奏；而選擇「荒涼沙漠」作為場景，則可以配合角色的孤獨表情，增強孤獨和絕望的情感表達。通過選擇和應用不同的場景，使用者可以更靈活地控制影像的情感氛圍，創造出具有深刻情感和視覺吸引力的作品。

以下我們提供了 100 種常用場景供創作者參考，此外，場景的描繪可以很單純的指出場所或地形地貌，但除此之外還可以加上更多細部的描述，例如添加物品、氣候、光線等等。

> **提示詞：一隻貓（A Cat）在（場景名稱）**

自然戶外

01. River
河邊

02. Forest
森林

03. Mountain top
山頂

04. Cave
山洞

05. Desert
沙漠

06. Field
田野

07. Pond
池塘

08. Waterfall
瀑布

09. Lake
湖泊

10. Canyon
峽谷

11. Volcano
火山

12. Glacier
冰川

13. Forest lake
森林湖

14. Wilderness
荒野

15. Beach
海灘

16. Stream
小溪

17. Gorge
峽谷

18. Island
島嶼

19. Peninsula
半島

20. Bay
海灣

城市居家

21. City
城市

22. Countryside
鄉村

23. Farm
農場

24. Hospital
醫院

25. School
學校

26. Library
圖書館

27. Restaurant
餐廳

28. Café
咖啡館

29. Supermarket
超市

30. Mall
商場

31. Airport
機場

32. Train station
火車站

33. Subway
地鐵

34. Bank
銀行

35. Bar
酒吧

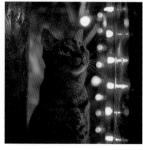

36. Nightclub
夜店

37. Theater
劇院

38. Museum
博物館

39. Art gallery
美術館

40. Stadium
體育場

休閒娛樂

41. Amuseme pntark
遊樂園

42.Zoo
動物園

43.Aquarium
水族館

44.Science center
科學館

45.Swimming pool
游泳池

46.Chalet
小木屋

47.Fishing village
漁村

48.Snow
雪地

49.Sandy beach
沙灘

50.Coral reef
礁石

51.Shrubbery
灌木叢

52.Botanical garden
植物園

53.Castle
城堡

54.Ruins
廢墟

55.Monastery
修道院

56.Church
教堂

57.Mosque
清真寺

58.Temple
廟宇

59.Cinema
電影院

60.Cabin
森林小屋

農業和鄉村

61. Ranch
牧場

62. Chicken coop
養雞場

63. Stable
馬廄

64. Warehouse
倉庫

65. Pier
碼頭

66. Dock
船塢

67. Boat
船上

68. Olive grove
橄欖園

69. Vineyard
葡萄園

70. Distillery
製酒廠

71. Granary
糧倉

72. Well
水井

73. Brook
溪流

74. Rainforest
熱帶雨林

75. Plain
曠野

76. Hunting ground
狩獵場

77. Flowerbed
花圃

78. Vegetable garden
菜園

79. Greenhouse
溫室

80. Fishing village
漁村

居家和建築

81. Cellar
地窖

82. Rooftop
屋頂

83. Attic
閣樓

84. Basement
地下室

85. Garage
車庫

86. Backyard
院子

87. Balcony
陽台

88. Terrace
露台

89. Rooftop garden
天台

90. Window
窗邊

91. Living room
客廳

92. Kitchen
廚房

93. Study
書房

94. Bedroom
臥室

95. Bathroom
浴室

96. Dining room
餐廳

97. Staircase
樓梯間

98. Porch
門廊

99. Garden
花園

100. Greenhouse
溫室

▶ 氛圍 🖥 272

在 Midjourney 中，除了視角定義，還有許多氛圍（Mood）Tokens，可以用來指定影像的整體情感和氛圍。這些 Tokens 有助於生成特定風格或感覺的影像。以下是一些常用的氛圍 Tokens，並附上範例提示詞。

01. Dark ｜黑暗

A dark, eerie forest at midnight, with thick fog and twisted trees.

02. Bright ｜明亮

A bright sunny day in a vibrant flower garden, with colorful blooms and butterflies.

03. Nostalgic ｜懷舊

A nostalgic scene of a 1950s diner, with vintage decor and a jukebox.

04. Romantic ｜浪漫

A romantic dinner setup by the beach at sunset, with candles and flowers.

05. Epic ｜史詩

An epic battle scene between knights and dragons in a vast medieval landscape.

06. Dreamy ｜夢幻

A dreamy, surreal landscape with floating islands and glowing flowers.

07. Serene ｜寧靜

A serene lakeside scene at dawn, with calm water and mist rising from the surface.

08. Futuristic ｜未來感

A futuristic cityscape at night, with towering skyscrapers and neon lights.

09. Desolate ｜荒涼

A desolate wasteland with abandoned buildings and cracked, dry earth.

10. 童話｜Fairytale

A fairytale forest with glowing mushrooms and a magical, sparkling stream.

11. Industrial ｜工業

An industrial factory interior, with large machinery and metal structures.

12. Festive ｜熱鬧

A festive street market at night, with colorful lights and bustling crowds.

13. Cold ｜冷峻

A cold, snowy landscape with a lone wolf standing on a frozen lake.

14. Sacred ｜神聖

A sacred temple interior, with golden statues and glowing candles.

15. Mysterious ｜神秘

A mysterious abandoned mansion, with overgrown vines and broken windows.

A dark, mysterious forest with a faint glow of moonlight filtering through the dense canopy. In the distance, an eerie, abandoned cabin emits a soft, nostalgic light.

這段提示詞結合了黑暗、神秘和懷舊氛圍，能夠生成一幅充滿情感層次的影像。

16. Wilderness ｜荒野

A wilderness scene with a dense forest, flowing river, and wildlife.

17. Quiet ｜安靜

A quiet, peaceful library with rows of old books and soft, ambient lighting.

18. Optimistic ｜樂觀

An optimistic sunrise over a blooming meadow, with birds chirping and a clear sky.

A futuristic, desolate landscape with towering, industrial ruins under a cold, overcast sky. In the distance, a fairytale-like glow emerges from a hidden forest, suggesting a sacred, magical refuge.

這段提示詞結合了未來感、荒涼、工業、冷峻、童話和神聖氛圍，能夠生成一幅豐富多層次且充滿對比的影像。

19. Adventurous ｜探險

An adventurous mountain trail with hikers, steep cliffs, and a breathtaking view.

20. Vibrant ｜充滿活力

A vibrant city street market with colorful stalls, lively crowds, and bright decorations.

21. Despair ｜絕望

A scene of despair with a lone figure standing in the rain, surrounded by dark, dilapidated buildings.

22. Magical ｜魔幻

A magical forest with glowing plants, mystical creatures, and sparkling lights.

23. Warm ｜溫暖

A warm, cozy living room with a fireplace, soft blankets, and a glowing lamp.

24. Inspirational ｜靈感

An inspirational mountain peak at sunrise, with a climber reaching the summit and raising their arms in triumph.

氛圍提示詞在 Midjourney 影像生成過程中起著至關重要的作用。這些詞彙允許使用者設定具體化影像的情感和氛圍，從而創造出具有特定感受和視覺效果的圖像。通過掌握和靈活運用各種氛圍，如黑暗、明亮、懷舊、浪漫、史詩、夢幻、寧靜、神秘等，使用者能夠精確地表達影像的內在情感和視覺意圖。

未來，隨著各類生成式平台模型以及語料、符元的迭代升級，我們可以期待更多樣化及精細定義的氛圍，進一步拓展影像生成的創作空間。只要能夠更好地掌握這些工具，創作出令人驚豔和富有情感的影像作品是件非常容易的事。

A vibrant, adventurous jungle scene with wild animals, magical glowing plants, and a group of explorers trekking through the dense vegetation under a warm, golden sunset.

屠龍

甲辰明荷篁東門書

PART 4

Midjourney
專屬風格定義

▶ 八大 Midjourney 專屬風格

在 Midjourney 早期推廣時，當時其創辦人 Holz David 在 Discord 的留言板上發布了一些有趣的風格關鍵字，這些內容雖然沒有系統化的被歸納分類，但它們是目前 Midjourney 平台唯一公布的官方風格樣式，如果能夠充分掌握這些風格特色，便可以用一個「Token」（符元），達到非常好的視覺效果。

接下來，將介紹「八大 Midjourney 專屬風格」，高達數百種的風格名稱或寫法，有時不一定具有正確的意義，可能是縮寫或是複合字，卻是系統可以辨認生成的符元。

範例

在核心風格裡有一個「Mommy's on-the-phonecore」（媽媽打電話？），看起來很像是工程師的幽默，但使用此風格可以快速生成具有母愛溫馨的畫面。

a dog and a cat in the style of Mommy's on-the-phonecore --ar 2:3 --style raw

想法 ⇒ 同樣一隻狗一隻貓，個別加上一個核心風格及龐克風格，看看會有甚麼樣的驚喜！

加上星際龐克風格。

a dog and a cat in the style of starpunk
（左）**Midjourney** 的生成效果
（右）**DALL.E** 的生成效果

加上彩虹核心風格。

a dog and a cat in the style of Rainbowcore
（左）**Midjourney** 的生成效果
（右）**DALL.E** 的生成效果

01. 核心風格（Core Styles） 🖥 283

01. Comfycore ｜ 舒適核心

圍繞舒適和溫暖的生活方式，色調柔和，通常帶有居家裝飾和放鬆氛圍。

cozy cat by a fireplace, comfycore

02. Cottagecore ｜ 田園核心

聚焦於鄉村生活和自然元素，色彩柔和，自然景觀和手工藝品是主要特徵。

A young woman in a quaint garden, cottagecore

03. Cranberrycore ｜ 蔓越莓核心

以蔓越莓色調為主，通常展示豐富的紅色和深紅色，具有濃厚的秋季和豐收氣氛。

forest scene with cranberry leaves, cranberrycore

04. 死亡核心 ｜ Deathcore

黑暗和激烈的風格，融合死亡金屬和哥特元素，色調多為黑色和深紅色。

dark knight in a gothic castle, deathcore

05. 夢幻核心 | Dreamcore

超現實迷幻風，強調夢境般的場景，色彩豐富且對比強烈，帶有神祕幻想。

surreal landscape with floating islands, dreamcore

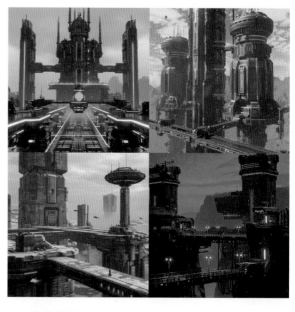

06. 終結者核心 | Endercore

靈感來自《安德的遊戲》，強調未來主義和軍事元素，色調冷峻。

futuristic military base, endercore

07. 故障核心 | glitchcore

聚焦數字和科技故障的美學，色調鮮明，常見失真和錯位的視覺效果。

digital glitch scene with bright visuals, glitchcore

08. 哥特核心 | Gothcore

結合哥特風格和現代元素，色調黑暗，強調神祕、浪漫和超自然的美學。

modern gothic castle with dark ambiance, gothcore

09. 陰森核心 | Grimcore

圍繞陰暗和悲慘的主題,色調非常暗沉,常見荒涼和破壞的景象。

desolate landscape with grim atmosphere, grimcore

10. 輾核核心 | Grindcore

強調極端速度和噪音的音樂風格,視覺上常見混亂和激烈的元素。

chaotic concert scene with intense energy, grindcore

11. 硬核核心 | Hardcore

音樂風格強調快速和激烈的節奏,視覺上常見強烈對比和動態場景。

intense mosh pit at a concert, hardcore

12. J 核核心 | Jcore

日本風格的硬核音樂,結合動畫和流行文化元素。

anime-themed concert with vibrant lights, jcore

13. Lovecore ｜愛心核心

浪漫和溫馨的視覺風格，色調多為粉紅色和紅色，強調愛情和幸福的元素。

couple holding hands in a rose garden, lovecore

14. Manticore ｜蠍獅核心

結合奇幻和神話元素，強調神祕和怪物的美學。

mythical creature in a dark forest, manticore

15. Neurocore ｜神經核心

聚焦於神經科學和未來科技，色調冷峻，強調高科技和未來感的元素。

futuristic lab with glowing neural networks, neurocore

16. Nightcore ｜夜核核心

快速而明亮的音樂風格，視覺上強調鮮豔色彩和快速移動的場景。

neon-lit dance club with high energy, nightcore

17. Pinkcore ｜粉紅核心

充滿粉紅色和可愛元素的風格，強調甜美和少女氣息。
a room decorated with pink balloons and plush toys, pinkcore

18. Polaroidcore ｜拍立得核心

模仿拍立得相片的視覺風格，強調懷舊和即時捕捉的感覺。
a candid moment captured on a polaroid, polaroidcore

19. Ragecore ｜憤怒核心

整體強調憤怒和力量的風格，色調激烈，視覺上充滿動感和力量。
a person screaming in the rain, ragecore

20. Schizocore ｜精神分裂核心

整體以混亂和分裂的視覺風格為主，強調心理錯亂和扭曲的元素。
a fragmented face with multiple expressions, schizocore

21. Stercore ｜排泄物核心

強調污垢和混亂的風格，色調暗沉，視覺上充滿不潔和混亂的場景。

a dirty alleyway with trash scattered around, stercore

22. Strifecore ｜衝突核心

強調衝突和對抗的風格，視覺上充滿張力和動態場景。

two warriors clashing swords in battle, strifecore

23. Traumacore ｜創傷核心

以創傷和悲傷為主題的風格，強調情感的表達和黑暗的視覺元素。

a person alone in a dark room with broken mirrors, traumacore

24. Webcore ｜網絡核心

整體聚焦於網絡和數字世界的風格，強調現代科技感和網絡元素。

a digital spider web with glowing lines, webcore

25. Weirdcore ｜ 怪異核心

以超現實和怪異的視覺風格為主，強調異常和反常的元素。
a bizarre landscape with floating eyes and strange shapes, weirdcore

26. Womancore ｜ 女性核心

強調女性氣質和力量的風格，視覺上充滿自信和力量的女性形象。
a powerful woman standing in front of a cityscape, womancore

27. Adventurecore ｜ 冒險核心

以冒險和探索為主題的風格，強調自然景觀和冒險精神。
a group of explorers climbing a mountain, Adventurecore

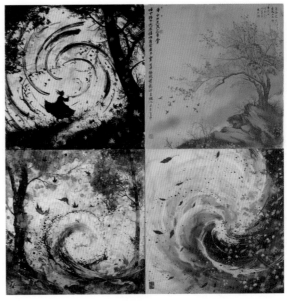

28. Anemoiacore ｜ 風神核心

整體以風神和風元素為主題的風格，強調動態的風景和自然力量。
a scene with swirling winds and flying leaves, Anemoiacore

29. Angelcore ｜天使核心

以天使和神聖元素為主題的風格，強調純潔和光明的視覺效果。

a glowing angel with white wings in a heavenly sky, Angelcore

30. Anglocore ｜英倫核心

聚焦於英國文化和風格，強調英倫風情和歷史元素。

a classic English manor house with lush gardens, Anglocore

31. Animecore ｜動漫核心

以日本動畫風格為主題，強調卡通和鮮豔的視覺元素。

a colorful anime character in a vibrant city, Animecore

32. Applecore ｜蘋果核心

以蘋果和果園為主題的風格，強調自然和農業元素。

a lush apple orchard with ripe apples on trees, Applecore

33. Auroracore ｜ 極光核心

聚焦於極光和北極景觀的風格，強調壯觀的自然現象。
a breathtaking aurora borealis over snowy mountains,
Auroracore

34. Babycore ｜ 寶寶核心

以嬰兒和幼兒元素為主題的風格，強調可愛和溫馨的氛圍。
a cozy nursery with soft toys and pastel colors,
Babycore

35. Barbiecore ｜ 芭比核心

整體以芭比娃娃和粉紅色調為主題的風格，強調華麗和夢幻元素。
a glamorous pink room with Barbie dolls and sparkles,
Barbiecore

36. Bardcore ｜ 吟遊詩人核心

以中世紀和民間傳說為主題的風格，強調古老的音樂和故事元素。
a medieval bard playing a lute by a campfire, Bardcore

37. Bastardcore │ 混蛋核心

粗暴和無賴風格,強調叛逆和挑釁的視覺元素。
a rebellious figure with a smirk, Bastardcore

38. Bloomcore │ 花開核心

以花朵和自然為主題的風格,強調繁榮和生命力。
a field of blooming wildflowers, Bloomcore

39. Bombacore │ 爆炸核心

充滿能量和動感的風格,強調爆炸和劇烈的視覺效果。
an action scene with explosions in the background,
Bombacore

40. Breakcore │ 碎核核心

強調高速和激烈的電子音樂風格,視覺上強調破碎和動感
的元素。
a digital artwork with fragmented shapes and bright
colors, Breakcore

41. Brocore ｜兄弟核心

主要強調兄弟情誼和友情的風格，通常包含活力和戶外活動元素。

a group of friends hiking in the mountains, Brocore

42. Bugcore ｜昆蟲核心

以昆蟲和自然為主題的風格，強調細節和自然界的奇妙。

a close-up of a colorful beetle on a leaf, Bugcore

43. Bunnycore ｜兔子核心

主要以兔子和可愛元素為主題的風格，強調溫馨和柔和的視覺效果。

a field with bunnies hopping around, Bunnycore

44. Cabincore ｜小屋核心

主要以森林小屋和自然生活為主題的風格，強調溫馨和自然元素。

a cozy cabin in the woods with a warm fireplace, Cabincore

45. Campcore ｜露營核心

聚焦於露營和戶外活動的風格，強調自然景觀和冒險精神。
a campsite by a lake with tents and a campfire,
Campcore

46. Candycore ｜糖果核心

充滿色彩繽紛和甜美元素的風格，強調糖果和甜點的視覺效果。
a whimsical scene with giant lollipops and candy
houses, Candycore

47. Caninecore ｜犬類核心

以狗和寵物為主題的風格，強調可愛和友好的氛圍。
a playful group of puppies in a park, Caninecore

48. Carcore ｜汽車核心

聚焦於汽車和機械元素的風格，強調速度和力量。
a sleek sports car on a racetrack, Carcore

49. Carnivalcore ｜嘉年華核心

充滿活力和色彩的風格，強調嘉年華和節慶元素。
a vibrant carnival scene with rides and games,
Carnivalcore

50. Cartelcore ｜黑幫核心

強調黑幫和犯罪元素的風格，視覺上充滿緊張和對抗。
a shadowy figure in a dark alleyway, Cartelcore

51. Cartooncore ｜卡通核心

以卡通和動畫為主題的風格，強調鮮豔色彩和誇張的視覺
元素。
a whimsical cartoon world with exaggerated characters,
Cartooncore

52. Catcore ｜貓核心

主要以貓和可愛元素為主題的風格，強調整體溫馨和愜意
的氛圍。
a cozy scene with cats lounging in the sun, Catcore

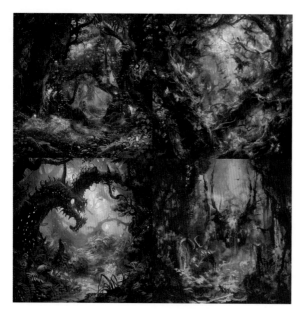

53. Changelingcore | 變形精靈核心

以奇幻和變形為主題的風格,強調神祕和魔法元素。
a magical forest with shapeshifting creatures,
Changelingcore

54. Christcore | 基督核心

聚焦於基督教和宗教元素的風格,強調神聖和精神的視覺效果。
a serene church scene with stained glass windows,
Christcore

55. Cleancore | 清潔核心

強調清潔和整潔的風格,視覺上充滿秩序和潔淨感。
a pristine white room with minimal decor, Cleancore

56. Cloudcore | 雲核心

以天空和雲朵為主題的風格,強調輕盈和夢幻的視覺效果。
a sky full of fluffy clouds, Cloudcore

57. Clowncore ｜小丑核心

充滿色彩和娛樂的風格，強調小丑和馬戲團元素。
a lively circus scene with clowns performing, Clowncore

58. Comiccore ｜漫畫核心

主要以漫畫和圖畫書為主題的風格，強調誇張和鮮明的視覺元素。
a superhero in a comic book style, Comiccore

59. Concore ｜會議核心

聚焦於商務和會議的風格，強調正式和專業的氛圍。
a business conference with people in suits, Concore

60. Craftcore ｜手工藝核心

以手工藝和創意為主題的風格，強調 DIY 和藝術創作。
a workshop filled with crafting materials, Craftcore

61. Crowcore ｜烏鴉核心

以烏鴉和黑暗元素為主題的風格，強調神祕和不祥的視覺效果。

a dark forest with crows perched on branches, Crowcore

62. Cryptidcore ｜神祕生物核心

聚焦於神祕生物和傳說生物的風格，強調奇幻和神祕元素。
a hidden creature lurking in the shadows, Cryptidcore

63. Crystalcore ｜水晶核心

以水晶和寶石為主題的風格，強調閃亮和華麗的視覺效果。
a cave filled with glowing crystals, Crystalcore

64. Cultcore ｜邪教核心

以邪教和儀式為主題的風格，強調神祕和黑暗的氛圍。
a dark ritual in an ancient temple, Cultcore

65. DarkErrorcore │ 黑暗錯誤核心

以數字和技術錯誤為主題的風格，強調黑暗和故障的視覺
效果。

a glitchy, dark digital landscape, DarkErrorcore

66. Dazecore │ 迷幻核心

聚焦於迷幻和夢境的風格，強調鮮豔色彩和超現實的視覺
效果。

a vibrant, surreal dreamscape, Dazecore

67. Devilcore │ 惡魔核心

主要是以惡魔和地獄為主題的風格，強調黑暗和恐怖的視
覺元素。

a demonic figure in a fiery underworld, Devilcore

68. Dinocore │ 恐龍核心

主要以恐龍和史前時代為主題的風格，強調古老和巨大的
視覺效果。

a prehistoric scene with dinosaurs roaming, Dinocore

69. Dollcore ｜娃娃核心

主要以娃娃和可愛元素為主題的風格，強調溫馨和柔和的視覺效果。

a room filled with vintage dolls, Dollcore

70. Dragoncore ｜龍核心

主要以龍和奇幻生物為主題的風格，強調壯觀和神祕的視覺效果。

a majestic dragon flying over mountains, Dragoncore

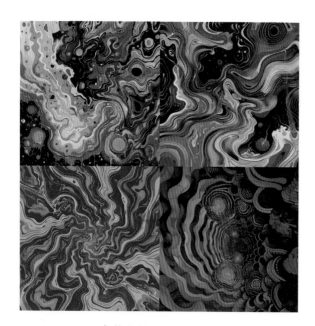

71. Drugcore ｜藥物核心

主要聚焦於藥物和迷幻體驗的風格，強調鮮豔色彩和扭曲的視覺效果。

a psychedelic scene with swirling patterns, Drugcore

72. Duckcore ｜鴨子核心

主要以鴨子和水禽為主題的風格，強調可愛和自然元素。

a pond with ducks swimming peacefully, Duckcore

73. Dullcore ｜乏味核心

強調單調和無趣的視覺風格，色調沉悶，強調日常生活的平凡。

a gray office space with repetitive tasks, Dullcore

74. Earthcore ｜地球核心

聚焦於地球和自然景觀的風格，強調環保和自然元素。

a lush forest scene with vibrant plant life, Earthcore

75. Emancicore ｜解放核心

以解放和自由為主題的風格，強調突破和變革的視覺元素。

a figure breaking free from chains, Emancicore

76. Fairycore ｜仙子核心

以仙子和奇幻為主題的風格，強調夢幻和魔法元素。

a magical fairy glen with sparkling lights, Fairycore

77. Fawncore │ 小鹿核心

主要以小鹿和森林生物為主題的風格,強調整體自然和溫馨的氛圍。

a fawn resting in a sunlit forest clearing, Fawncore

78. Feralcore │ 野性核心

聚焦於野生和未經馴化的風格,強調整體自然和原始的視覺效果。

a wild animal prowling through a dense jungle, Feralcore

79. Firecore │ 火焰核心

以火焰和熱力為主題的風格,強調燃燒和動感的視覺效果。

a roaring bonfire with sparks flying, Firecore

80. Frogcore │ 青蛙核心

以青蛙和濕地生物為主題的風格,強調自然和可愛的元素。

a pond with lily pads and frogs leaping, Frogcore

81. Gamercore ｜遊戲核心

聚焦於遊戲和電子競技的風格，強調數字和整體動感的視覺效果。

a high-tech gaming setup with neon lights, Gamercore

82. Ghostcore ｜幽靈核心

以幽靈和超自然現象為主題的風格，強調神祕和恐怖的視覺效果。

a haunted house with ghostly apparitions, Ghostcore

83. Gloomcore ｜陰鬱核心

強調黑暗和憂鬱的風格，色調陰暗，視覺上充滿沉重和壓抑的氛圍。

a dark, rainy cityscape with a melancholic mood, Gloomcore

84. Goblincore ｜哥布林核心

聚焦於哥布林和奇幻生物的風格，強調奇特和神祕的視覺效果。

a goblin village hidden in a forest, Goblincore

85. Gorpcore ｜戶外核心

聚焦於戶外活動和露營的風格，強調實用和自然元素。
a group of hikers with camping gear in a mountain landscape, Gorpcore

86. Grandparentcore ｜祖父母核心

強調懷舊和家庭溫暖的風格，充滿溫馨和回憶的元素。
a cozy living room with vintage decor and family photos, Grandparentcore

87. Groundcore ｜地面核心

以大地和自然為主題的風格，強調穩定和自然元素。
a serene landscape with earthy tones and grounded elements, Groundcore

88. Halloweencore ｜萬聖節核心

以萬聖節和恐怖主題的風格，強調神祕和恐怖的視覺效果。
a spooky haunted house with jack-o'-lanterns, Halloweencore

89. Happycore │ 快樂核心

以快樂和積極為主題的風格,強調明亮和活力的效果。
a sunny meadow with people laughing and playing,
Happycore

90. Hatecore │ 仇恨核心

強調憤怒和對抗的風格,視覺上充滿強烈和尖銳的元素。
a riot scene with intense emotions and confrontations,
Hatecore

91. Havencore │ 避風港核心

以安全和舒適為主題的風格,強調寧靜和庇護的氛圍。
a peaceful sanctuary with soft lighting and comfortable
seating, Havencore

92. Heistcore │ 搶劫核心

聚焦於搶劫和犯罪冒險的風格,強調緊張和計劃的效果。
a thrilling heist scene with masked characters and
vaults, Heistcore

93. Hikecore ｜ 徒步核心

以徒步旅行和探險為主題的風格，強調自然和探索元素。
a scenic trail with hikers and breathtaking views, Hikecore

94. Hispanicore ｜ 西班牙核心

聚焦於西班牙文化和傳統的風格，強調熱情和色彩繽紛的視覺效果。
a vibrant Spanish fiesta with traditional costumes and dances, Hispanicore

95. Honeycore ｜ 蜂蜜核心

以蜂蜜和蜜蜂為主題的風格，強調自然和甜美的元素。
a sunny field with buzzing bees and honeycombs, Honeycore

96. Intel Core ｜ 英特爾核心

聚焦於現代科技和計算機元素的風格，強調高科技感。
a sleek computer lab with advanced technology, Intel Core

97. Jamcore │果醬核心

以果醬和甜美為主題的風格,強調家庭和手工製作的溫馨氛圍。

a rustic kitchen with jars of homemade jam, Jamcore

98. Junglecore │叢林核心

以叢林和熱帶雨林為主題的風格,強調豐富和多樣的自然景觀。

a dense jungle with exotic plants and animals, Junglecore

99. Karencore │凱倫核心

以現代社會和社交媒體為主題的風格,強調諷刺和現實。

a dramatic scene with a stereotypical "Karen" character, Karencore

100. Kidcore │兒童核心

聚焦於童年和玩具的風格,強調明亮的色彩和遊戲元素。

a colorful playroom filled with toys, Kidcore

101. Kimoicore ｜奇萌核心

結合了奇幻和可愛元素的風格，強調夢幻和萌系視覺效果。
a magical land with adorable creatures, Kimoicore

102. Kingcore ｜國王核心

以國王和皇室為主題的風格，強調華麗和尊貴的元素。
a grand throne room with a majestic king, Kingcore

103. Knightcore ｜騎士核心

聚焦於中世紀和騎士的風格，強調勇敢和榮譽。
va valiant knight in shining armor, Knightcore

104. Kuromicore ｜黑魅核心

以黑暗和神祕為主題的風格，強調哥特和奇幻元素。
a dark forest with mysterious shadows, Kuromicore

105. Labcore │ 實驗室核心

聚焦於實驗室和科學研究的風格,強調高科技和創新。
a futuristic lab with scientists at work, Labcore

106. LI-core │ LI 核心

聚焦於線性插畫和簡潔的風格,強調簡約和清晰的效果。
a minimalist illustration with clean lines, LI-core

107. Maidcore │ 女僕核心

以女僕和服務為主題的風格,強調溫馨和細緻的視覺元素。
a quaint cafe with maids serving tea, Maidcore

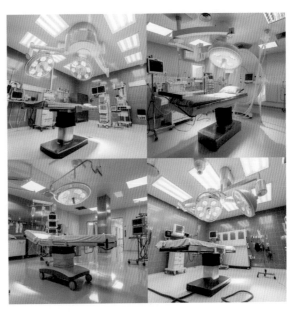

108. Medicalcore │ 醫療核心

聚焦於醫療和健康的風格,強調專業和潔淨的視覺效果。
a modern hospital room with medical equipment,
Medicalcore

109. Metalcore ｜ 金屬核心

以金屬音樂和工業元素為主題的風格，強調黑暗和強烈的視覺效果。

a heavy metal concert with intense lighting, Metalcore

110. Miniaturecore ｜ 微型核心

以微型物品和場景為主題的風格，強調細節和精緻的視覺效果。

a tiny, detailed diorama of a city, Miniaturecore

111. Mommy's on-the-phonecore ｜ 媽媽打電話核心

以家庭和日常生活為主題的風格，強調溫馨和現實的元素。

a mom multitasking at home, Mommy's on-the-phonecore

112. Mosscore ｜ 苔蘚核心

聚焦於苔蘚和森林景觀的風格，強調自然和濕潤的效果。

a lush forest floor covered in moss, Mosscore

113. Mushroomcore ｜蘑菇核心

以蘑菇和菌類為主題的風格，強調奇幻和自然的元素。
a whimsical forest with giant mushrooms, Mushroomcore

114. Naturecore ｜自然核心

聚焦於自然和戶外景觀的風格，強調生態和環保的效果。
a pristine forest with diverse wildlife, Naturecore

115. Nerdcore ｜書呆子核心

以書呆子文化和愛好為主題的風格，強調智力和興趣愛好。
a cozy room filled with books and gadgets, Nerdcore

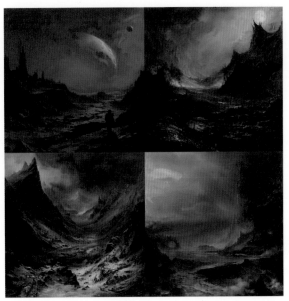

116. Nihilcore ｜虛無核心

強調虛無主義和黑暗哲學的風格，視覺上充滿沉思和陰暗的元素。
a bleak landscape with an empty void, Nihilcore

117. Nintencore ｜任天堂核心

聚焦於任天堂遊戲和角色的風格，強調懷舊和樂趣的視覺效果。

a colorful scene with iconic Nintendo characters, Nintencore

118. Normcore ｜常態核心

以平凡和日常生活為主題的風格，強調簡單和現實的視覺效果。

a typical suburban street with everyday scenes, Normcore

119. Nostalgiacore ｜懷舊核心

聚焦於懷舊和過去的風格，強調溫暖和回憶的視覺效果。
a vintage room with old photos and memorabilia, Nostalgiacore

120. Old Memecore ｜老梗核心

聚焦於過去的網絡迷因和經典梗的風格，強調懷舊和幽默。
a scene featuring classic internet memes, Old Memecore

121. Paleocore ｜ 古生物核心

以古生物和史前時代為主題的風格，強調古老和巨大的視覺效果。

a prehistoric scene with dinosaurs and ancient landscapes, Paleocore

122. Petcore ｜ 寵物核心

聚焦於寵物和可愛動物的風格，強調整體溫馨和可愛的視覺效果。

a cozy home with various pets playing together, Petcore

123. Pigeoncore ｜ 鴿子核心

以鴿子和城市鳥類為主題的風格，強調日常和城市生活的元素。

a bustling city scene with pigeons flying around, Pigeoncore

124. Poetcore ｜ 詩人核心

聚焦於詩歌和文學的風格，強調浪漫和知性的視覺效果。

a serene study with an old-fashioned writing desk and books, Poetcore

125. Poolcore ｜ 游泳池核心

整體以游泳池和水上活動為主題的風格，強調夏日和休閒的氛圍。

a vibrant swimming pool scene with people relaxing and playing, Poolcore

126. Prairiecore ｜ 草原核心

聚焦於草原和農村生活的風格，強調大自然和簡樸生活的視覺效果。

a wide open prairie with wildflowers and rustic farmhouses, Prairiecore

127. Prehistoricore ｜ 史前核心

以史前時代和古代文化為主題的風格，強調古老和神祕的元素。

an ancient cave with primitive drawings and artifacts, Prehistoricore

128. Princecore ｜ 王子核心

聚焦於王子和皇室生活的風格，整體強調華麗和浪漫的視覺效果。

a royal palace with a charming prince, Princecore

129. Princesscore ｜公主核心

以公主和童話為主題的風格，強調華麗和夢幻的元素。
a fairy tale castle with a beautiful princess, Princesscore

130. Queencore ｜女王核心

聚焦於女王和權力的風格，強調華麗和尊貴的視覺效果。
a majestic throne room with a regal queen, Queencore

131. Queercore ｜酷兒核心

以酷兒文化和多樣性為主題的風格，強調色彩繽紛和自我
表現。
a vibrant pride parade with colorful outfits, Queercore

132. Rainbowcore ｜彩虹核心

聚焦於彩虹和色彩的風格，強調明亮和愉快的視覺效果。
a bright, rainbow-colored landscape with cheerful
elements, Rainbowcore

133. Rangercore │ 巡護員核心

以自然保護和巡護員為主題的風格，強調戶外和保護自然的元素。
a forest ranger patrolling a lush forest, Rangercore

134. Ratcore │ 老鼠核心

整體以老鼠和城市動物為主題的風格，強調都市生活和生存元素。
a bustling city alley with rats scurrying about, Ratcore

135. Ravencore │ 烏鴉核心

以烏鴉和神祕為主題的風格，強調黑暗和不祥的視覺效果。
a dark forest with ravens perched on branches, Ravencore

136. Retrocore │ 復古核心

聚焦於過去幾十年的風格，強調懷舊和經典元素。
a vintage 80s room with old-school electronics, Retrocore

137. Roguecore ｜流浪者核心

以流浪和自由為主題的風格，強調冒險和獨立的元素。
a lone traveler on a rugged path, Roguecore

138. Rotcore ｜腐爛核心

聚焦於腐爛和廢墟的風格，強調黑暗和荒涼的視覺效果。
a decaying urban environment with crumbling buildings, Rotcore

139. Royalcore ｜皇家核心

以皇室和貴族生活為主題的風格，強調華麗和尊貴的視覺效果。
a grand palace with opulent decorations, Royalcore

140. Rural China ｜農村中國

聚焦於中國農村生活和文化的風格，強調自然和傳統元素。
a peaceful rural village in China with traditional houses, Rural China

141. Rusticcore ｜鄉村核心

以鄉村生活和自然材料為主題的風格，強調簡樸和自然的
視覺效果。

a cozy farmhouse with wooden furniture, Rusticcore

142. SacriCore ｜神聖核心

聚焦於宗教和神聖儀式的風格，整體強調神祕和崇高的視
覺效果。

a solemn cathedral with candles and stained glass,
SacriCore

143. Sanriocore ｜三麗鷗核心

以三麗鷗角色和可愛元素為主題的風格，強調色彩繽紛和
萌系視覺效果。

a colorful scene with animals and friends, Sanriocore

144. Scoutcore ｜童軍核心

聚焦於童軍活動和戶外探險的風格，強調自然和團隊精神。

a group of scouts around a campfire in the woods,
Scoutcore

145. Scrapbook │剪貼簿核心

以剪貼簿和手工藝為主題的風格，強調創意和懷舊的元素。
a detailed scrapbook with various mementos and photos, Scrapbook

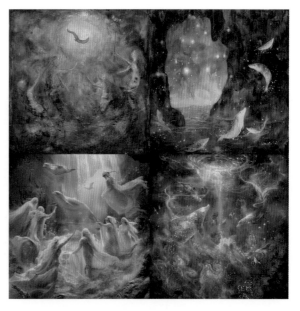

146. Selkiecore │海豹人核心

聚焦於海豹人傳說和海洋元素的風格，強調神祕和奇幻的視覺效果。
a mystical sea with selkies transforming, Selkiecore

147. Sleepycore │睡意核心

以睡眠和放鬆為主題的風格，強調溫馨和安靜的氛圍。
a cozy bedroom with soft pillows and blankets, Sleepycore

148. Smilecore │微笑核心

聚焦於快樂和積極情緒的風格，強調明亮和愉快的效果。
a cheerful scene with people smiling and laughing, Smilecore

149. Snailcore ｜蝸牛核心

以蝸牛和慢生活為主題的風格，強調緩慢和悠閒的氛圍。
a serene garden with snails leisurely moving, Snailcore

150. Soggycore ｜濕潤核心

聚焦於濕潤和潮濕環境的風格，強調水和潮濕的元素。
a rainy forest with lush greenery, Soggy

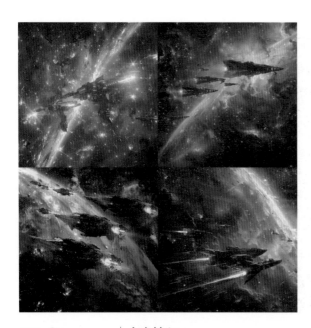

151. Spacecore ｜太空核心

以太空和科幻為主題的風格，強調廣闊和未來感的效果。
a starry galaxy with futuristic spaceships, Spacecore

152. Sparklecore ｜閃耀核心

聚焦於閃耀和華麗的風格，強調閃光和華麗的視覺效果。
a dazzling scene with sparkling lights and glitter,
Sparklecore

153. Spiritcore │ 靈魂核心

整體以靈魂和精神世界為主題的風格,強調神祕和超自然的元素。

a mystical forest with ethereal spirits, Spiritcore

154. Spiritualcore │ 靈性核心

聚焦於靈性和冥想的主題風格,整體強調寧靜和內心的視覺效果。

a serene temple with people meditating, Spiritualcore

155. Stripcore │ 極簡核心

以剝離和簡約為主題的風格,強調極簡和現代的視覺效果。

a minimalist room with bare essentials, Stripcore

156. Technocore │ 科技核心

聚焦於現代科技和未來主義的風格,強調高科技和數字化的視覺效果。

a high-tech cityscape with neon lights and advanced technology, Technocore

157. Teethcore ｜牙齒核心

以牙齒和牙科為主題的風格，強調獨特和有趣的視覺效果。
a quirky dental office with oversized teeth decor,
Teethcore

158. Terrorwave ｜恐怖核心

以恐怖和驚悚為主題的風格，強調懸疑和恐怖的視覺效果。
a chilling scene with eerie shadows and mist,
Terrorwave

159. Thriftcore ｜節儉核心

聚焦於節儉和二手店的風格，強調懷舊和經濟實惠的視覺
效果。
a cozy thrift store with vintage finds, Thriftcore

160. Tinkercore ｜修補核心

以修補和 DIY 為主題的風格，強調創意和實用的視覺效果。
a workshop filled with tools and gadgets, Tinkercore

161. Tinycore | 微型核心

聚焦於小型物品和微縮世界的風格，強調精緻和細緻的視覺效果。

a miniature village with tiny houses and figures, Tinycore

162. Toycore | 玩具核心

以玩具和童年回憶為主題的風格，強調色彩繽紛和樂趣的視覺效果。

a vibrant toy store with a variety of toys, Toycore

163. Traincore | 火車核心

聚焦於火車和鐵路元素的風格，整體強調懷舊和旅行的視覺效果。

a classic steam train chugging through the countryside, Traincore

164. Transportcore | 運輸核心

以各種運輸方式為主題的風格，整體強調動感和機械的視覺效果。

a bustling transport hub with various vehicles, Transportcore

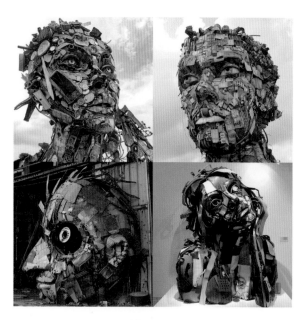

165. Trashcore ｜垃圾核心

聚焦於廢棄物和回收利用的風格，強調環保和創意的視覺效果。

a creative sculpture made from recycled materials, Trashcore

166. Trenchcore ｜戰壕核心

以戰壕和軍事歷史為主題的風格，強調歷史和嚴肅的視覺效果。

a World War I trench scene with soldiers, Trenchcore

167. Trendercore ｜潮流核心

聚焦於時尚潮流和當代風格的風格，強調時尚和創新的視覺效果。

a trendy fashion runway with the latest styles, Trendercore

168. Tweencore ｜青少年核心

以青少年文化和興趣為主題的風格，強調活力和年輕的視覺效果。

a lively school scene with teens hanging out, Tweencore

169. Unicorncore ｜ 獨角獸核心

以獨角獸和奇幻元素為主題的風格，強調夢幻和神祕的視覺效果。

a magical forest with unicorns grazing, Unicorncore

170. Urbancore ｜ 都市核心

聚焦於城市生活和現代建築的風格，強調繁忙和現代的視覺效果。

a bustling cityscape with skyscrapers and busy streets, Urbancore

171. Vacation Dadcore ｜ 度假老爸核心

以家庭度假和輕鬆氛圍為主題的風格，強調愉快和休閒的視覺效果。

a dad in a Hawaiian shirt enjoying a beach vacation, Vacation Dadcore

172. Villagecore ｜ 村莊核心

聚焦於村莊生活和鄉村景觀的風格，強調自然和簡樸的視覺效果。

a quaint village with cobblestone streets and cottages, Villagecore

173. Voidcore │ 虛無核心

強調虛無和空無的風格，視覺上充滿抽象和簡約的元素。
a vast empty space with minimalistic design, Voidcore

174. Warmcore │ 溫暖核心

以溫暖和舒適為主題的風格，強調溫馨和愜意的氛圍。
a cozy living room with a roaring fireplace, Warmcore

175. Weathercore │ 天氣核心

聚焦於天氣和氣象現象的風格，強調自然和變化的效果。
a dramatic thunderstorm with lightning and rain,
Weathercore

176. Wetcore │ 潮濕核心

以濕潤和水元素為主題的風格，強調水和潮濕的視覺效果。
a rainy day with puddles and reflections, Wetcore

177. Witchcore 　│巫術核心

聚焦於巫術和魔法元素的風格，強調神祕和奇幻的效果。
a dark forest with a witch casting spells, Witchcore

178. Wizardcore 　│巫師核心

以巫師和魔法為主題的風格，強調神祕和學術的視覺效果。
a grand library with a wise old wizard, Wizardcore

179. Wormcore 　│蠕蟲核心

整體以蠕蟲和地下生物為主題的風格，強調自然和細緻的
視覺效果。
a garden soil scene with worms and roots, Wormcore

180. Yankeecore 　│洋基核心

聚焦於美國東北部文化和歷史的風格，整體強調歷史和傳
統元素。
a classic New England town with colonial architecture,
Yankeecore

181. Zombiecore ｜殭屍核心

以殭屍和末日為主題的風格，強調恐怖和生存的視覺效果。
a post-apocalyptic world with zombies roaming,
Zombiecore

▶ 02. 龐克風格（Punk Styles）🖥 328

01. Aetherclockpunk ｜以太鐘龐克

結合了以太與蒸汽龐克的元素，通常呈現複雜機械結構和
華麗齒輪裝置的場景，色調多為古銅色、金色和褐色。
a steampunk engineer working on a fantastical machine
with gears and cogs, aetherclockpunk

02. Aetherpunk ｜以太龐克

融合科幻和奇幻元素，以太能量驅動的未來世界，畫面充
滿光影效果，強調神祕和先進技術的結合，色調通常是藍
色、紫色和銀色。
a scientist studying a glowing orb in a futuristic city
powered by aether energy, aetherpunk

03. Algeapunk ｜藻類龐克

以海洋和水生植物為主題，展示水下世界的奇幻景象。色彩多為綠色和藍色，充滿自然和神祕感。

a diver exploring an underwater city with glowing algae and marine creatures, algeapunk

04. Alienpunk ｜外星龐克

整體以外星生物和星際旅行為主題，強調未知和神祕的宇宙元素。

various alien species interacting in a bustling marketplace on a distant planet, alienpunk

05. Atompunk ｜原子龐克

聚焦於 20 世紀中期的原子能科技，強調復古未來主義和科技感。

scientists in vintage lab coats working in a retro-futuristic laboratory with atomic gadgets, atompunk

06. Aurorapunk ｜極光龐克

以極光和極地景觀為主題，強調壯觀的自然現象和神祕感。

explorers in thermal gear admiring a stunning arctic landscape with colorful auroras, aurorapunk

07. Autopunk │自動龐克

整體以自動化和機械裝置為主題，強調工業和機械化的視覺效果。
workers operating intricate automated machines in a factory, autopunk

08. Avocadopunk │酪梨龐克

聚焦於酪梨和健康生活方式，強調自然和有機元素。
a chef preparing gourmet avocado dishes in a stylish kitchen, avocadopunk

09. Berrypunk │漿果龐克

以各種漿果為主題，強調豐富的色彩和自然元素。
a gardener tending to a lush garden filled with various berries, berrypunk

10. Biopunk │生物龐克

以生物科技和基因工程為主題，強調科幻和未來生物技術的視覺效果。
a scientist conducting experiments on genetically engineered creatures in a futuristic lab, biopunk

11. Carpetpunk ｜地毯龐克

以地毯和紡織品為主題，強調復古和手工藝的視覺效果。
an artisan weaving intricate designs on a traditional loom, carpetpunk

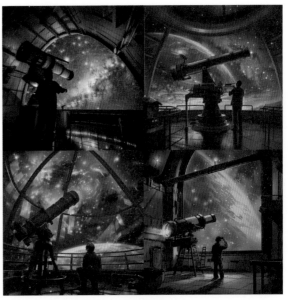

12. Celestialpunk ｜天體龐克

聚焦於天體和宇宙元素，強調壯觀和神祕的視覺效果。
an astronomer observing glowing stars through a high-tech telescope in a celestial observatory, celestialpunk

13. Chinapunk ｜中國龐克

以中國傳統文化和未來元素為主題，強調古今融合的視覺效果。
a futuristic city with traditional Chinese architecture and modern technology, chinapunk

14. Chromepunk ｜鉻龐克

聚焦於金屬和反光表面，強調未來感和科技感的視覺效果。
a sleek, metallic cityscape with chrome surfaces and futuristic vehicles, chromepunk

15. Chronopunk ｜時間龐克

整體以時間旅行和歷史變遷為主題，強調複雜和動態的視覺效果。

time travelers navigating shifting historical settings with futuristic devices, chronopunk

16. Citypunk ｜城市龐克

聚焦於現代都市生活和未來都市景觀，強調繁忙和動感的視覺效果。

a bustling futuristic city with skyscrapers and neon lights, people going about their daily lives, citypunk

17. Clockpunk ｜時鐘龐克

以時鐘和齒輪裝置為主題，強調機械和工藝的視覺效果。

a clockmaker crafting intricate clockwork mechanisms in a workshop, clockpunk

18. Cloudpunk ｜雲龐克

聚焦於天空和雲層，強調輕盈和夢幻的視覺效果。

a floating city above the clouds with citizens and flying creatures, cloudpunk

19. Clownpunk ｜小丑龐克

以小丑和馬戲團為主題，強調色彩繽紛和古怪的視覺效果。
a vibrant circus with clowns and acrobats performing,
clownpunk

20. Coralpunk ｜珊瑚龐克

聚焦於海洋和珊瑚礁生態系，強調鮮豔色彩和自然的美麗。
a marine biologist studying vibrant coral reefs with
colorful fish, coralpunk

21. Cottagepunk ｜田園龐克

結合了田園風和未來元素，強調自然與科技的結合。
a futuristic farmer tending to robotic animals in a
countryside cottage, cottagepunk

22. Crustpunk ｜硬皮龐克

聚焦於地下音樂和街頭文化，強調粗糙和反叛的視覺效果。
punk rockers performing on a gritty street corner with
graffiti-covered walls, crustpunk

23. Cryptopunk ｜加密龐克

聚焦於加密技術和數字貨幣，強調高科技和未來金融的視覺效果。

a hacker in a digital city with holographic cryptocurrency symbols, cryptopunk

24. Cyberaetherpunk ｜網絡以太龐克

整體結合了網絡科技與以太元素，強調未來感和神祕的視覺效果。

cybernetic beings interacting in a digital landscape with ethereal data streams, cyberaetherpunk

25. Cybermysticpunk ｜網絡神祕龐克

聚焦於網絡世界和神祕元素的結合，強調神祕和未來感。

futuristic monks in a temple with mystical symbols and holographic guardians, cybermysticpunk

26. Cyberpunk ｜網絡龐克

聚焦於高科技和反烏托邦未來的視覺風格，強調黑暗和先進技術。

a neon-lit cityscape with cybernetic humans and flying cars, cyberpunk

27. Cyberraypunk ｜網絡射線龐克

聚焦於光線和激光科技，強調明亮和未來感的視覺效果。
soldiers in futuristic armor engaged in a laser battle in a high-tech city, cyberraypunk

28. Cybersteampunk ｜網絡蒸汽龐克

整體結合了蒸汽龐克去和網絡科技的元素，強調復古和未來的融合。
a steampunk city with cybernetic enhancements and steam-powered machinery, cybersteampunk

29. Cyperpunk ｜網絡龐克

聚焦於網絡和數字世界的風格，強調現代科技和網絡元素。
a digital city with hackers and cybernetic beings interacting, cyperpunk

30. Cypherpunk ｜密碼龐克

聚焦於加密和安全技術的風格，強調數字隱私和科技感。
a secretive hacker decrypting data streams in a high-tech lab, cypherpunk

31. Decopunk ｜裝飾藝術龐克

結合了裝飾藝術和科技元素，強調華麗和復古未來主義。
people dressed in art deco fashion walking through a
glamorous futuristic city, decopunk

32. Derppunk ｜蠢龐克

聚焦於滑稽和荒誕的視覺風格，強調搞笑和輕鬆的元素。
comical characters in a whimsical city with exaggerated
features, derppunk

33. Desertpunk ｜沙漠龐克

聚焦於沙漠和荒地的風格，強調荒涼和生存的元素。
a scavenger with futuristic gear wandering through a
desolate desert landscape, desertpunk

34. Dieselpunk ｜柴油龐克

聚焦於柴油科技和工業革命的風格，強調粗糙和工業化的
視覺效果。
a gritty mechanic working on diesel-powered machines
in an industrial city, dieselpunk

35. Draftpunk ｜草稿龐克

聚焦於設計草稿和未完成作品的風格，強調創意和未來感。
artists sketching futuristic designs in a high-tech studio,
draftpunk

36. Dreampunk ｜夢幻龐克

結合夢幻和未來科技元素，強調超現實和迷幻的效果。
children playing in a surreal landscape with floating
islands and ethereal beings, dreampunk

37. Fantasypunk ｜奇幻龐克

聚焦於奇幻世界和科技元素的結合，強調魔法和先進技術
的融合。
wizards using magical technology in a fantastical city
with mythical creatures, fantasypunk

38. Fiberpunk ｜纖維龐克

聚焦於纖維和紡織科技的風格，強調細緻和未來感。
fashion designers creating intricate textiles with futuristic
looms, fiberpunk

39. Floralpunk ｜ 花卉龐克

結合了花卉和未來科技元素，強調自然和人工技術的融合。
a botanist tending to genetically enhanced flowers in a futuristic greenhouse, floralpunk

40. Foampunk ｜ 泡沫龐克

聚焦於泡沫和流體科技的風格，強調流動和透明的效果。
scientists experimenting with glowing foam and liquid in a futuristic lab, foampunk

41. Fractalpunk ｜ 分形龐克

結合了分形幾何和未來科技的元素，強調複雜和對稱的視覺效果。
architects designing fractal buildings in a symmetrical futuristic city, fractalpunk

42. Frostpunk ｜ 霜凍龐克

聚焦於冰雪和極地科技的風格，強調寒冷和未來感。
explorers in advanced thermal suits navigating an icy landscape, frostpunk

43. Futurepunk ｜未來龐克

聚焦於未來科技和超現實世界的風格,強調前衛和創新的視覺效果。

a family living in a sleek, futuristic home with advanced gadgets, futurepunk

44. Genepunk ｜基因龐克

聚焦於基因工程和生物科技的風格,強調科幻和未來生物技術的視覺效果。

scientists working on genetically engineered animals in a high-tech lab, genepunk

45. Geopunk ｜地質龐克

聚焦於地質學和地球科技的風格,強調自然和科學的結合。

geologists using advanced equipment to study rock formations in a futuristic landscape, geopunk

46. Ghoulpunk ｜鬼魂龐克

聚焦於鬼魂和超自然元素,強調恐怖和神祕的視覺效果。

ghostly figures haunting a futuristic city with eerie lighting, ghoulpunk

47. Glitchpunk │故障龐克

聚焦於數字和科技故障的美學，色調鮮明，常見失真和錯位的視覺效果。
a hacker surrounded by glitchy visuals and digital distortions, glitchpunk

48. Goosepunk │鵝龐克

聚焦於鵝和鄉村生活，強調自然和輕鬆的氛圍。
a farmer herding geese in a peaceful countryside, goosepunk

49. Gothicpunk │哥德龐克

結合哥德風格和現代元素，色調黑暗，強調神祕、浪漫和超自然的美學。
a vampire in a gothic cathedral with futuristic elements, gothicpunk

50. Icepunk │冰龐克

聚焦於冰雪和極地科技的風格，強調寒冷和未來感。
polar explorers in advanced thermal gear navigating icy landscapes, icepunk

51. Industrialpunk ｜工業龐克

聚焦於工業革命和現代科技的風格，強調粗糙和工業化的
視覺效果。
factory workers operating heavy machinery in an
industrial landscape, industrialpunk

52. Iunglepunk ｜叢林龐克

聚焦於叢林和熱帶生態系，強調自然和冒險的元素。
a biologist studying exotic plants and animals in a dense
jungle, junglepunk

53. Kawaiipunk ｜可愛龐克

結合可愛元素和龐克風格，強調色彩繽紛和萌系視覺效果。
characters in colorful outfits with cute accessories in a
vibrant city, kawaiipunk

54. Kombuchapunk ｜康普茶龐克

聚焦於康普茶和健康生活方式，強調自然和有機元素。
customers enjoying kombucha in a trendy café with
organic decor, kombuchapunk

55. Magipunk ｜魔法龐克

結合魔法和科技元素，強調奇幻和先進技術的融合。
wizards using magical devices in a technologically advanced city, magipunk

56. Manapunk ｜法力龐克

聚焦於魔法和法力的元素，強調神祕和奇幻的視覺效果。
sorcerers harnessing mana in a mystical city with advanced technology, manapunk

57. Mcdonaldpunk ｜麥當勞龐克

結合麥當勞和快餐文化的元素，整體強調現代和諷刺的視覺效果。
fast food workers in a dystopian cityscape with robotic servers, mcdonaldpunk

58. Meatpunk ｜肉類龐克

聚焦於肉類和生物科技的元素，強調科幻和未來生物技術的視覺效果。
butchers handling bioengineered meats in a futuristic market, meatpunk

59. Necropunk │ 死靈龐克

聚焦於死亡和死靈法術的元素,整體強調黑暗和恐怖的視覺效果。

necromancers raising the dead in a haunted futuristic city, necropunk

60. Neonpunk │ 霓虹龐克

聚焦於霓虹燈和城市夜景的元素,強調色彩鮮豔和未來感的視覺效果。

city dwellers walking through neon-lit streets with futuristic signs, neonpunk

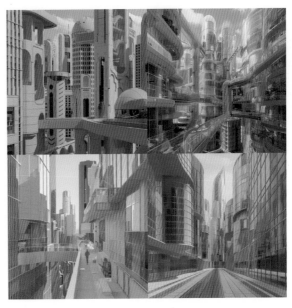

61. Neopunk │ 新龐克

結合現代和未來元素的風格,強調創新和前衛的視覺效果。

urbanites in a sleek, modern city with cutting-edge technology, neopunk

62. Pantonepunk │ Panton 龐克

聚焦於潘通色卡和色彩的元素,強調多彩和視覺沖擊力。

designers creating vibrant, Pantone-colored buildings in a futuristic city, pantonepunk

63. Piratepunk ｜海盜龐克

結合海盜和現代科技的元素，強調冒險和自由的視覺效果。
futuristic pirates sailing high-tech ships on open seas,
piratepunk

64. Poodlepunk ｜貴賓犬龐克

聚焦於貴賓犬和可愛元素的結合，強調時尚和趣味。
stylish individuals walking their poodles in a whimsical
park, poodlepunk

65. Punk ｜龐克

聚焦於傳統龐克文化和音樂的元素，強調反叛和自我表現。
punk rockers with mohawks and leather jackets
performing on a gritty street, punk

66. Quantumpunk ｜量子龐克

結合量子物理和科技元素，強調複雜和超現實的視覺效果。
scientists experimenting with quantum technology in a
futuristic lab, quantumpunk

67. Ragepunk │ 憤怒龐克

聚焦於憤怒和力量的元素，強調強烈和動感的視覺效果。
warriors battling in a chaotic, dystopian city with intense energy, ragepunk

68. Seapunk │ 海洋龐克

聚焦於海洋和水生生物的元素，強調自然和神祕感。
divers exploring an underwater city filled with colorful marine life, seapunk

69. Shitpunk │ 垃圾龐克

聚焦於廢棄物和回收利用的元素，強調環保和創意的視覺效果。
artists creating sculptures from recycled materials in an urban junkyard, shitpunk

70. Sithpunk │ 西斯龐克

結合了《星際大戰》中西斯和龐克元素，強調黑暗和強大的視覺效果。
Sith Lords wielding lightsabers in a dark, futuristic cityscape, sithpunk

71. Slaughterpunk │ 屠宰龐克

聚焦於屠宰和肉類加工的元素，整體強調黑暗和恐怖的視覺效果。

butchers in a grim slaughterhouse with advanced technology, slaughterpunk

72. Solarpunk │ 太陽能龐克

同樣聚焦於太陽能和可再生能源的元素，整體強調環保和未來感。

citizens living in a solar-powered city with green technology, solarpunk

73. Spacepunk │ 太空龐克

聚焦於太空和星際旅行的元素，整體強調廣闊和神祕的視覺效果。

astronauts exploring a futuristic space station with advanced spacecraft, spacepunk

74. Spacesolarpunk │ 太空太陽能龐克

結合太空和太陽能元素，強調可持續能源和未來科技。

astronauts working on solar panels in a space colony, spacesolarpunk

75. Starpunk ｜星際龐克

聚焦於星際旅行和外太空的元素，強調探索和冒險的視覺效果。

a starship captain navigating through distant galaxies, starpunk

76. Steampunk ｜蒸汽龐克

結合了 19 世紀工業革命和未來科技的元素，強調蒸汽動力和復古未來主義。

Victorian gentlemen operating steam-powered machines, steampunk

77. Steempunk ｜煉鋼龐克

聚焦於煉鋼和工業科技的元素，強調工業化和力量的視覺效果。

workers in a steel mill forging futuristic machinery, steempunk

78. Synthpunk ｜合成龐克

結合合成音樂和未來科技的元素，強調電子和未來感的視覺效果。

musicians playing synthesizers in a neon-lit club, synthpunk

79. Techpunk │ 科技龐克

同樣聚焦於現代科技和數字世界的元素，強調高科技和未來感。

programmers coding advanced AI in a futuristic office, techpunk

80. Vaporpunk │ 蒸汽船龐克

結合了蒸汽船和未來科技的元素，強調復古和未來感的融合。

inventors working on steam-powered airships in a futuristic city, vaporpunk

81. Xmaspunk │ 聖誕龐克

聚焦於聖誕節和節日元素的結合，強調歡樂和節慶的視覺效果。

people celebrating Christmas with steampunk decorations and gadgets, xmaspunk

82. Zuckerpunk │ 祖克龐克

結合科技和企業文化的元素，整體強調現代和高科技的視覺效果。

corporate employees using advanced tech in a futuristic office, zuckerpunk

83. Bibliopunk ｜書籍龐克

聚焦於書籍和文學的元素，強調知識和神祕的視覺效果。
scholars reading ancient tomes in a high-tech library, Bibliopunk

84. Bronzepunk ｜青銅龐克

結合青銅時代和科技元素，強調古代和未來感的融合。
artisans crafting bronze statues with futuristic tools, Bronzepunk

85. Cargopunk ｜貨運龐克

聚焦於貨運和物流的元素，強調工業和未來感的視覺效果。
workers loading advanced cargo ships in a futuristic port, Cargopunk

86. Corporate Punk ｜企業龐克

結合企業文化和反叛元素，強調現代和諷刺的視覺效果。
rebels disrupting a high-tech corporate meeting, Corporate Punk

87. Cripplepunk 殘疾龐克

聚焦於殘疾和科技的元素，強調力量和未來感。
individuals with advanced prosthetics navigating a
futuristic city, Cripplepunk

88. CyberneticPunk ｜賽博龐克

結合了賽博技術和龐克元素，強調高科技和未來感的視覺
效果。
cybernetic beings interacting in a neon-lit city,
CyberneticPunk

89. Daydreampunk ｜白日夢龐克

聚焦於夢幻和幻想元素的結合，強調超現實和奇幻的效果。
children exploring a dreamlike landscape with fantastical
creatures, Daydreampunk

90. Dinopunk ｜恐龍龐克

結合恐龍和現代科技元素，強調古代和未來感的融合。
adventurers riding robotic dinosaurs in a futuristic
jungle, Dinopunk

91. Dracopunk ｜龍龐克

聚焦於龍和奇幻元素的結合，強調壯觀和神祕的視覺效果。
dragon riders navigating a city with futuristic architecture, Dracopunk

92. Europunk ｜歐洲龐克

結合歐洲文化和龐克元素，強調多元和現代的視覺效果。
Europeans in punk fashion exploring historic and futuristic landmarks, Europunk

93. Folk Punk ｜民俗龐克

結合民俗文化和龐克元素，強調傳統和現代的融合。
villagers in traditional attire with punk accessories celebrating a festival, Folk Punk

94. Forestpunk ｜森林龐克

聚焦於森林和自然元素的結合，強調綠色和環保的效果。
forest dwellers using advanced eco-friendly technology in a lush forest, Forestpunk

95. Gadgetpunk ｜小工具龐克

聚焦於小工具和科技元素的結合，強調創新和實用的視覺效果。

inventors showcasing their advanced gadgets in a tech fair, Gadgetpunk

96. Hermitpunk ｜隱士龐克

聚焦於隱士生活和自然元素的結合，強調孤獨和寧靜的視覺效果。

hermits living in secluded cabins with advanced technology, Hermitpunk

97. Lunarpunk ｜月球龐克

結合月球和未來科技元素，強調神祕和超現實的視覺效果。
astronauts building a futuristic moon base, Lunarpunk

98. Mythpunk ｜神話龐克

結合神話和現代科技元素，強調奇幻和先進技術的融合。
demigods using advanced technology in a mythological setting, Mythpunk

99. Nanopunk | 納米龐克

聚焦於納米科技和未來科技元素的結合，強調微觀和超現實的視覺效果。
scientists manipulating nanobots in a futuristic lab, Nanopunk

100. Pastel Punk | 粉彩龐克

結合粉彩和龐克元素，強調色彩鮮豔和輕鬆的視覺效果。
teenagers in pastel punk fashion enjoying a vibrant cityscape, Pastel Punk

101. Post-Punk | 後龐克

聚焦於後龐克音樂和文化的元素，強調黑暗和冷酷的視覺效果。
musicians performing in a dark, moody venue with post-punk aesthetics, Post-Punk

102. Salvagepunk | 廢料龐克

聚焦於廢料和回收利用的元素，整體強調環保和創意的視覺效果。
artists creating sculptures from recycled materials in an urban junkyard, Salvagepunk

103. Sandalpunk | 涼鞋龐克

結合涼鞋和現代科技元素，強調舒適和創新的視覺效果。
beachgoers in futuristic sandals enjoying a high-tech beachfront, Sandalpunk

104. Slimepunk | 黏液龐克

聚焦於黏液和生物科技的元素，強調科幻和未來生物技術的視覺效果。
scientists experimenting with glowing slime in a futuristic lab, Slimepunk

105. Steelpunk | 鋼鐵龐克

聚焦於鋼鐵和工業科技的元素，強調力量和未來感的視覺效果。
workers building futuristic steel structures in an industrial city, Steelpunk

106. Stonepunk | 石器龐克

結合石器時代和現代科技元素，強調古代和未來感的融合。
prehistoric humans using advanced stone technology in a futuristic setting, Stonepunk

107. Swordpunk ｜劍龐克

聚焦於劍和中世紀戰鬥的元素，整體強調力量和榮譽的視覺效果。

knights with glowing swords battling in a futuristic arena, Swordpunk

108. Tupinipunk ｜土著龐克

結合土著文化和現代科技元素，強調傳統和現代的融合。

villagers using advanced tools in a traditional settlement, Tupinipunk

109. Vaporwavepunk ｜蒸汽波龐克

結合蒸汽波和龐克元素，強調懷舊和未來感的融合。

characters in retro fashion with futuristic elements in a neon-lit city, vaporwavepunk

110. Venuspunk ｜金星龐克

聚焦於金星和科幻元素的結合，整體強調神祕和外星的視覺效果。

colonists exploring Venus with advanced technology, venuspunk

111. Victorianpunk │ 維多利亞龐克

結合維多利亞時代和現代科技元素,強調古典和未來感的融合。
individuals in Victorian attire with futuristic gadgets in a steampunk city, victorianpunk

112. Vikingspunk │ 維京龐克

聚焦於維京文化和現代科技元素的結合,強調力量和冒險的視覺效果。
Viking warriors using advanced weapons in a futuristic setting, vikingspunk

113. Waterpunk │ 水龐克

聚焦於水和海洋科技元素,強調自然和未來感的視覺效果。
marine biologists using advanced equipment in an underwater city, waterpunk

114. Werewolfpunk │ 狼人龐克

結合狼人和現代科技元素,強調黑暗和神祕的視覺效果。
werewolves with cybernetic enhancements in a dark, futuristic city, werewolfpunk

115. Wildwestpunk │荒野西部龐克

結合荒野西部和現代科技元素，整體強調冒險和自由的視覺效果。

cowboys using advanced technology in a futuristic Wild West town, wildwestpunk

116. Witchpunk │巫師龐克

聚焦於巫師和魔法元素的結合，整體強調神祕和奇幻的視覺效果。

witches casting spells with advanced technology in a mystical setting, witchpunk

117. Woodpunk │木頭龐克

結合木材和現代科技元素，強調自然和環保的視覺效果。

carpenters crafting advanced wooden structures in a futuristic forest, woodpunk

118. Xenomorphpunk │異形龐克

聚焦於異形和科幻元素的結合，強調恐怖和未來感的視覺效果。

humans encountering xenomorphs in a dark, futuristic setting, xenomorphpunk

119. Y2kpunk ｜千禧年龐克

聚焦於 2000 年和現代科技元素的結合，強調懷舊和未來感
的融合。

characters in Y2K fashion using futuristic technology,
y2kpunk

120. Zeppelinpunk ｜齊柏林龐克

結合齊柏林飛船和現代科技元素，整體強調懷舊和未來感
的融合。

pilots flying zeppelin airships with advanced technology,
zeppelinpunk

121. Zombiepunk ｜殭屍龐克

聚焦於殭屍和末日元素的結合，整體強調恐怖和生存的視
覺效果。

survivors battling zombies in a post-apocalyptic world
with advanced weapons, zombiepunk

03. 藝術風格（Art Styles） 🖥 359

01. 3D Art │ 三度藝術

結合了數位技術和三度空間的藝術形式，整體強調深度和立體感。
a 3D sculpture of a futuristic cityscape with animated characters, 3D Art

02. Abstract Art │ 抽象藝術

強調形狀、色彩和線條的表達，而非具象的描繪，注重情感和思想的表達。
an abstract painting with vibrant colors and geometric shapes, Abstract Art

03. Abstract Expressionism │ 抽象表現主義

注重自發性和即興創作，強調情感的強烈表達和自由的藝術形式。
a dynamic and expressive painting with bold brushstrokes, Abstract Expressionism

04. Anamorphic Art │ 變形藝術

利用透視原理創作的扭曲影像，從特定角度觀看可恢復正常形象。
an anamorphic drawing that reveals a hidden image from a specific angle, Anamorphic art

05. Appropriation Art ｜ 挪用藝術

使用現有的影像或物品重新創作，賦予其新的意義和文脈。
a collage of famous artworks reinterpreted in a modern context, Appropriation art

06. Art Deco ｜ 裝飾藝術

結合幾何圖形和奢華裝飾，強調現代感和精緻美學，常見於建築和設計中。
a luxurious building with geometric patterns and rich decorations, Art Deco

07. Art Informel ｜ 非形式藝術

強調自發性和非傳統技法，追求抽象和表現主義的結合。
a spontaneous and abstract painting with freeform shapes and textures, Art Informel

08. Art Nouveau ｜ 新藝術

主要結合自然形態和曲線美學，常見於建築、設計和裝飾藝術中。
a building with flowing, organic lines and floral motifs, Art Nouveau

09. Arte Povera ｜窮人藝術

使用簡單和日常材料創作，強調自然和社會主題。
a sculpture made from everyday objects and natural materials, Arte Povera

10. Calligraphy Art ｜書法藝術

以優美的線條和字形創作，強調書寫的藝術性和表達力。
an intricate calligraphy piece with elegant strokes and flowing script, Calligraphy

11. Color Field ｜色域繪畫

注重大面積單一或漸變色塊的運用，強調色彩的純粹性和視覺效果。
a large canvas with bold, uniform color fields and subtle gradients, Color Field

12. Conceptual Art ｜概念藝術

強調藝術背後的概念和思想，而非具體的物質形式。
an installation that challenges traditional notions of art and perception, Conceptual Art

13. Concrete Art ｜ 具體藝術

強調藝術作品的物質性和現實性，排除任何抽象或象徵性的成分。

a geometric sculpture with precise lines and solid forms, Concrete Art

14. Constructivism ｜ 構成主義

強調機械化和技術創新，注重幾何結構和功能性的結合。

a structure made from industrial materials and geometric shapes, Constructivism

15. Contemporary Realism ｜ 當代寫實主義

注重精細和逼真的描繪，結合現代主題和技法。

a highly detailed painting of a modern urban scene, Contemporary Realism

16. Cubism ｜ 立體主義

分解和重組視覺元素，強調多視角和幾何形態的結合。

a fragmented and geometric painting of a still life, Cubism

17. Culture Jamming │ 文化干擾

使用媒體和大眾文化元素創作，強調批判和諷刺社會現象。
a provocative artwork that subverts popular
advertisements and logos, Culture Jamming

18. Cynical Realism │ 憤世嫉俗的寫實主義

起源於中國，結合幽默和諷刺，反映現實和社會問題。
a satirical painting depicting modern social issues with
exaggerated realism, Cynical Realism

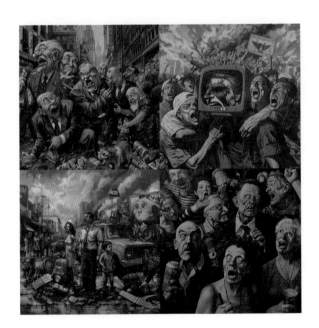

19. Dada │ 達達主義

否定傳統藝術形式和美學，強調荒謬和無序。
a chaotic and nonsensical collage that defies artistic
conventions, Dada

20. Doodle Art │ 塗鴉藝術

以即興和自由的形式創作，強調創意和個性。
a playful and intricate doodle filled with whimsical
characters and patterns, Doodle Art

21. Drip Painting ｜滴畫

以滴灑顏料的方式創作，強調動感和隨機性。
an artist creating a dynamic piece by dripping vibrant colors onto a canvas, Drip Painting

22. Environmental Art ｜環境藝術

與自然環境互動創作，強調環保和生態意識。
a large-scale installation using natural materials in a forest, Environmental Art

23. Expressionism ｜表現主義

強調主觀情感和內心世界的表達，常用強烈色彩和扭曲形象詮釋。
a painter expressing intense emotions with bold colors and distorted figures, Expressionism

24. Fantasy Art ｜奇幻藝術

描繪神話和幻想世界，充滿想像力和超自然元素。
a fantastical scene with dragons, wizards, and magical landscapes, Fantasy Art

25. Fauvism ｜野獸派

使用強烈、非自然的色彩和簡化的形象表達情感。
a vibrant painting with bold, unrealistic colors and simple
forms, Fauvism

26. Feminist Art ｜女性主義藝術

強調女性經歷和社會地位，探討性別議題。
a powerful piece addressing gender issues and
celebrating women's strength, Feminist Art

27. Figuration narrative ｜敘事具象藝術

結合敘事性和具象表現，講述故事或反映社會現實。
a detailed painting telling a complex story with realistic
characters, Figuration narrative

28. Figurative Art ｜具象藝術

聚焦於現實世界中的形象和物體，強調真實性和細節。
a lifelike portrait capturing the essence of a person,
Figurative Art

29. Figurative Expressionism ｜具象表現主義

結合具象和表現主義元素，強調情感和形象的表達。
a portrait with exaggerated features conveying deep emotion, Figurative Expressionism

30. Fluxus ｜激浪派

強調藝術和生活的結合，倡導即興創作和多媒體藝術。
a mixed-media performance blending visual art, music, and theater, Fluxus

31. Funk art ｜放克藝術

結合流行文化和非正統材料，強調幽默和反叛精神。
a whimsical sculpture using found objects and bright colors, Funk art

32. Futurism ｜未來主義

強調速度、科技和現代生活，表現動感和力量。
a dynamic painting depicting futuristic machines and movement, Futurism

33. Geometric abstraction ｜幾何抽象藝術

使用幾何形狀和純粹色彩創作，強調結構和形式。
an abstract piece with precise geometric shapes and
bold colors, Geometric abstraction

34. Graffiti ｜塗鴉藝術

在公共空間創作的街頭藝術，強調自發性和反叛精神。
a colorful mural on an urban wall, featuring bold letters
and characters, Graffiti

35. Gutai group ｜具體派

日本戰後前衛藝術團體，強調物質和行動的結合。
an interactive performance involving physical
engagement with materials, Gutai group

36. Hard-edge Painting ｜硬邊繪畫

強調清晰的邊界和純粹色塊，追求簡潔和明確性。
a painting with distinct, sharp-edged shapes and vibrant
colors, Hard-edge Painting

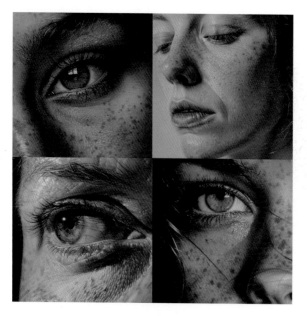

37. Hyperrealism ｜超寫實主義

追求極致的細節和真實感，超越攝影效果。
a highly detailed painting or sculpture that looks almost like a photograph, Hyperrealism

38. Illustration ｜插畫

用於書籍、雜誌等媒材的圖像創作，強調敘事性和視覺吸引力。
a whimsical illustration for a children's book, featuring colorful characters and scenes, Illustration

39. Imagism ｜意象派

詩歌和藝術中的一個流派，強調清晰的形象和精煉的表達。
a minimalist painting capturing a vivid image with simple, clear lines, Imagism

40. Impressionism ｜印象派

追求光影和色彩的瞬間效果，常用短暫筆觸和自然主題。
a serene landscape with soft, diffused light and vibrant colors, Impressionism

41. Kinetic Art ｜動力藝術

以運動和動態效果為特點的藝術形式，常用機械裝置。
a sculpture with moving parts that create a dynamic visual experience, Kinetic Art

42. Kitsch ｜刻奇藝術

刻意模仿和誇大流行文化，強調誇張和幽默。
a playful piece that exaggerates popular cultural elements in a humorous way, Kitsch

43. Land Art ｜大地藝術

在自然環境中創作的大型戶外作品，強調環境和地景。
a large-scale earthwork that transforms the natural landscape, Land Art

44. Les Nabis ｜那比派

19 世紀末的法國藝術團體，強調裝飾性和象徵性。
a decorative painting with symbolic imagery and vibrant colors, Les Nabis

45. Lettrism │字母主義

以字母和文字為主要表現元素的藝術流派,強調語言和視覺的結合。
a text-based artwork that explores the visual aspects of letters and words, Lettrism

46. Light art │光藝術

使用光源和照明效果創作的藝術形式,強調視覺和氛圍。
an installation using colored lights to create a captivating atmosphere, Light art

47. Lyrical Abstraction │抒情抽象

強調自發性和表現力的抽象藝術形式,注重色彩和筆觸的情感表達。
a vibrant abstract painting with flowing lines and dynamic brushstrokes, Lyrical Abstraction

48. Magic Realism │魔幻現實主義

結合現實和幻想元素,創造出神祕而真實的視覺效果。
a painting that blends realistic details with fantastical elements, Magic Realism

49. Mail art ｜ 郵寄藝術

通過郵寄方式創作和傳遞的藝術形式，強調交流和合作。
an art piece made of postcards and letters sent through the mail, Mail art

50. Massurrealism ｜ 大眾超現實主義

結合超現實主義和大眾文化元素，創造出夢幻和現實的奇異融合。
a surreal artwork that incorporates elements of pop culture and everyday life, Massurrealism

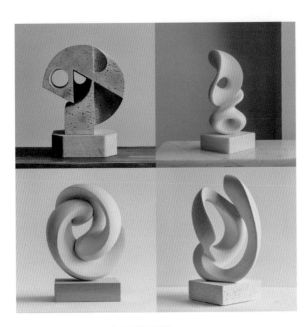

51. Minimal Art ｜ 極簡藝術

強調簡約和形式的純粹性，通常使用單純的形狀和色彩。
a minimalist sculpture with clean lines and simple forms, Minimal Art

52. Modernism ｜ 現代主義

強調創新和打破傳統，探索新的藝術形式和技術。
an abstract painting that breaks traditional conventions with innovative techniques, Modernism

53. Monochromatic Art ｜ 單色藝術

使用單一色調創作，強調色彩的微妙變化和視覺效果。
a painting in shades of blue that explores different tones and depths, Monochromatic Art

54. Murals ｜ 壁畫

大型公共藝術作品，通常在建築物牆面上創作，強調社區和文化的表達。
a vibrant mural on a city wall depicting local culture and history, Murals

55. Naïve Art ｜ 純真藝術

由非專業藝術家創作，強調簡單和直接的表現方式。
a colorful and whimsical painting by a self-taught artist, Naïve art

56. Neo-Conceptualism ｜ 新概念藝術

強調觀念和思維的創作形式，探索藝術的意義和社會問題。
an installation that challenges traditional art concepts with thought-provoking themes, Neo-Conceptualism

57. Neo-Dada │新達達

結合達達主義的反叛精神和現代元素,強調荒謬和挑釁。
a mixed-media artwork that combines absurdity and modern elements, Neo-Dada

58. Neo-Expressionism │新表現主義

強調情感和個人表達,使用粗獷的筆觸和強烈的色彩。
a dynamic painting with bold brushstrokes and vibrant colors, Neo-Expressionism

59. Neo-figurative art │新具象藝術

結合具象和現代元素,強調細節和現實性。
a detailed portrait with contemporary themes and realistic features, Neo-figurative art

60. Neo-Impressionism │新印象派

結合印象派的光影效果和現代技法,強調視覺效果。
a landscape painting that captures light and color with modern techniques, Neo-Impressionism

61. Neo-Pop ｜新波普

結合波普藝術和現代文化元素，整體強調色彩鮮豔和視覺衝擊力。
a vibrant artwork that blends pop culture icons with modern aesthetics, Neo-Pop

62. Neo-primitivism ｜新原始主義

結合原始藝術和現代元素，強調簡單和樸素的表現方式。
a sculpture that incorporates primitive forms and modern techniques, Neo-primitivism

63. Neo-romanticism ｜新浪漫主義

強調浪漫和情感表達，結合現代和古典元素。
a dreamy landscape with romantic themes and modern twists, Neo-romanticism

64. Neue Wilde ｜新狂野派

20 世紀 70 年代的德國藝術運動，整體強調自發性和強烈的色彩。
a painting with bold colors and spontaneous brushstrokes, Neue Wilde

65. New European Painting ｜新歐洲繪畫

結合歐洲傳統和現代技法，強調細節和表達力。
a contemporary European painting with detailed and expressive forms, New European Painting

66. New Leipzig School ｜新萊比錫學派

起源於德國萊比錫，結合具象和抽象元素，強調社會和個人主題。
a figurative painting with abstract elements and social themes, New Leipzig School

67. Nouveau Réalisme ｜新寫實主義

1960 年代的法國藝術運動，強調現實和社會主題的表現。
an artwork that uses everyday objects to depict social realities, Nouveau Réalisme

68. Op Art ｜光學藝術

利用視覺錯覺創作，強調動感和視覺效果。
an optical illusion artwork that creates a sense of movement and depth, Op Art

69. Orphism ｜奧菲主義

結合色彩和音樂的藝術形式，強調感官體驗。
a colorful abstract painting that evokes musical rhythms and harmonies, Orphism

70. Outsider Art ｜外行藝術

由非專業藝術家創作，強調個人和自發性。
an expressive and unconventional artwork by a self-taught artist, Outsider Art

71. Photorealism ｜照相寫實主義

以高度細節和精確度模仿照片效果的藝術形式。
a painting that looks indistinguishable from a high-resolution photograph, Photorealism

72. Plastic arts ｜塑膠藝術

使用塑膠和合成材料創作，強調創新和材料的多樣性。
a sculpture made from brightly colored plastic materials, Plastic arts

73. Pointillism │ 點描畫

使用小點來構成圖像，強調色彩和光影效果。
a landscape painting created with thousands of tiny dots of color, Pointillism

74. Political Pop │ 政治波普

結合波普藝術和政治主題，強調社會批判和諷刺。
an artwork that combines pop culture imagery with political messages, Political Pop

75. Pop Art │ 波普藝術

結合大眾文化和日常物品，強調色彩和視覺吸引力。
a vibrant artwork featuring famous pop culture icons and bright colors, Pop Art

76. Pop Surrealism │ 波普超現實主義

結合波普藝術和超現實主義元素，強調幻想和夢幻效果。
a whimsical and surreal painting with pop culture influences, Pop Surrealism

77. Post-Impressionism │後印象派

在印象派基礎上發展出的藝術流派，強調個性和創新。
a painting that blends impressionist techniques with unique, personal style, Post-Impressionism

78. Post-painterly Abstraction │後繪畫抽象主義

20 世紀 60 年代的藝術運動，強調色彩和形狀的純粹性。
an abstract artwork with clean lines and pure color fields, Post-painterly Abstraction

79. Postmodernism │後現代主義

批判現代主義，強調多元和不確定性。
an eclectic artwork that combines various styles and media, Postmodernism

80. Primitivism │原始主義

模仿原始和土著藝術，強調簡單和自然的表現方式。
a sculpture inspired by tribal art with simple, raw forms, Primitivism

81. Process art ｜過程藝術

注重創作過程而非最終結果，強調動作和材料的變化。
an artist creating a piece through repetitive and
meditative actions, Process art

82. Pseudorealism ｜擬真主義

強調細節和逼真效果，但包含超現實或幻想元素。
a hyper-realistic painting with surreal twists,
Pseudorealism

83. Psychedelic art ｜迷幻藝術

使用鮮豔色彩和扭曲形象，反映迷幻體驗和視覺錯覺。
a vibrant and colorful painting with swirling patterns and
shapes, Psychedelic art

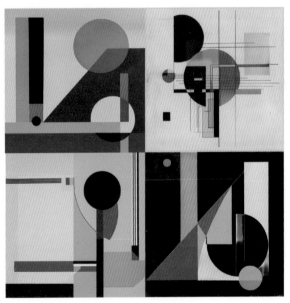

84. Purism ｜純粹主義

強調簡單和秩序，使用簡化的形狀和線條。
an artwork with clean lines and simple geometric
shapes, Purism

85. Realism ｜寫實主義

注重細節和真實性，真實描繪現實生活。
a detailed portrait that captures the essence of a real person, Realism

86. Romanticism ｜浪漫主義

強調情感和個人表達，常用自然和歷史主題。
a dramatic landscape painting with romantic and emotional themes, Romanticism

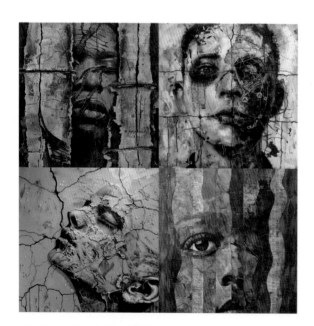

87. Scar Art ｜疤痕藝術

使用疤痕和傷口為主題，強調身體和心理創傷的表達。
an artwork that incorporates scar motifs to tell a story of healing, Scar Art

88. Social Realism ｜社會寫實主義

注重社會問題和現實，反映工人和普通人的生活。
a painting depicting the daily lives of working-class individuals, Social Realism

89. Stuckism ｜反觀念主義

批判當代藝術，強調具象繪畫和個人表達。
a figurative painting with bold and expressive forms,
Stuckism

90. Superflat ｜超扁平主義

結合日本流行文化和傳統藝術，強調平面和簡化。
a vibrant artwork featuring flat, colorful characters
inspired by anime, Superflat

91. Suprematism ｜至上主義

強調基本幾何形狀和純粹色彩，追求純粹抽象。
an abstract painting with basic geometric forms and
pure colors, Suprematism

92. Surrealism ｜超現實主義

結合現實和夢幻元素，創造出奇異和超現實的景象。
a dreamlike scene with bizarre and fantastical elements,
Surrealism

93. Symbolism │ 象徵主義

使用象徵性圖像表達內心世界和思想，常用神話和夢幻題
材來詮釋。

a mystical painting with symbolic figures and themes,
Symbolism

94. Synchromism │ 同步主義

強調色彩的和諧和音樂性的表現方式，使用色彩創造韻律
和節奏。

an abstract painting that uses harmonious colors to
evoke a sense of rhythm, Synchromism

95. Tachisme │ 塗鴉派

強調自發性和隨意的色彩運用，注重動感和表現力。

a dynamic abstract painting with splashes and drips of
vibrant colors, Tachisme

96. Tape Art │ 膠帶藝術

使用膠帶創作的藝術形式，強調創意和臨時性。

an intricate mural made entirely of colored tape, Tape
Art

97. Transavantgarde ｜超前衛藝術

結合不同藝術風格和媒材，強調多樣性和創新。
an eclectic artwork that blends various styles and media, Transavantgarde

98. Ukiyo-e ｜浮世繪

17 至 19 世紀的日本木版畫風格，常描繪風景、歌舞伎演員和美人。
a traditional Japanese woodblock print depicting a scenic landscape, Ukiyo-e

99. Urban Art ｜城市藝術

在城市環境中創作的藝術形式，包括有塗鴉、壁畫和裝置藝術。
a large-scale mural on an urban building, featuring vibrant street art, Urban Art

100. Wildstyle ｜狂野風格

塗鴉藝術的一種，特點是複雜和錯綜的字體設計。
a graffiti piece with intricate and interwoven letterforms, Wildstyle

101. Yarn Bombing │毛線炸彈

在公共場所用毛線包裹和裝飾物體的藝術形式，強調溫暖和人性化。
a tree covered in colorful knitted yarn, bringing warmth to an urban environment, Yarn Bombing

102. Young British Artists │英國青年藝術家

1990 年代興起的英國藝術家群體，作品常見具有挑釁性和實驗性。
a provocative and experimental piece by a young British artist, Young British Artists

103. Zero Movement │零運動

1950 年代的歐洲藝術運動，強調光、運動和空間的表現。
an installation using light and movement to create dynamic visual effects, Zero Movement

01. Acidwave ｜酸波風潮

結合了迷幻和電子音樂的元素，強調鮮豔色彩和視覺錯覺的效果。
a vibrant neon cityscape with swirling patterns and psychedelic characters, Acidwave

02. Desertwave ｜沙漠風潮

結合沙漠景觀和電子音樂元素，強調荒涼和神祕感。
a futuristic desert with glowing sand dunes and electronic structures, Desertwave

03. Fashwave ｜時尚風潮

結合時尚和電子音樂元素，強調前衛和華麗的視覺效果。
a high-fashion runway show with neon lights and futuristic outfits, Fashwave

04. Glowwave ｜光風潮

結合發光效果和電子音樂元素，強調光影和視覺沖擊。
a glowing forest with luminescent plants and animals, Glowwave

05. Hallyu │韓流

結合韓國流行文化和電子音樂元素，強調現代和流行的視覺效果。

a bustling street in Seoul with K-pop billboards and neon signs, Hallyu

06. Heatwave │熱浪風潮

結合高溫和電子音樂元素，強調熱烈和動感的視覺效果。

a scorching urban landscape with heatwaves and vibrant street scenes, Heatwave

07. Hustlewave │奔波風潮

結合都市生活和電子音樂的元素，強調生活忙碌和現代的視覺效果。

busy city streets with people in motion and digital billboards, Hustlewave

08. Laborwave │勞動風潮

結合勞動文化和電子音樂元素，整體強調工人階級和現代社會的視覺效果。

factory workers in a futuristic setting with neon lights and machinery, Laborwave

09. Lightningwave ｜ 閃電風潮

結合閃電和電子音樂元素，強調動感和能量的視覺效果。
a futuristic cityscape illuminated by lightning and neon lights, Lightningwave

10. Magewave ｜ 魔法風潮

結合魔法和電子音樂元素，強調奇幻和神祕的視覺效果。
wizards casting spells in a neon-lit fantasy city, Magewave

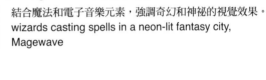

11. Milleniwave ｜ 千禧風潮

結合千禧一代文化和電子音樂元素，強調懷舊和現代感。
a cityscape featuring elements from the 2000s with futuristic enhancements, Milleniwave

12. New Wave ｜ 新潮

結合 20 世紀 70-80 年代的新潮音樂和現代元素，強調復古和未來感。
people dancing to new wave music in a neon-lit club, New Wave

13. Palewave │ 淡色風潮

結合淡色調和電子音樂元素，強調柔和和舒緩的視覺效果。
a serene landscape with pastel colors and soft lighting,
Palewave

14. Reefwave │ 珊瑚礁風潮

結合海洋和電子音樂元素，強調自然和動感的視覺效果。
an underwater scene with vibrant coral reefs and
luminescent sea creatures, Reefwave

15. Rollerwave │ 滾軸風潮

結合滾軸溜冰和電子音樂元素，整體強調動感和復古的視
覺效果。
people roller skating in a neon-lit rink with futuristic
decorations, Rollerwave

16. Schizowave │ 精神分裂風潮

結合混亂和多變的電子音樂元素，強調不穩定和複雜的視
覺效果。
a chaotic cityscape with rapidly changing colors and
patterns, Schizowave

17. Sovietwave ｜蘇聯波風潮

結合蘇聯時代的美學和電子音樂元素，強調懷舊和獨特的
視覺效果。
a retro-futuristic city with Soviet architecture and neon
lights, Sovietwave

18. Stimwave ｜刺激風潮

結合高刺激的電子音樂元素，強調動感和能量的視覺效果。
an energetic urban scene with vibrant lights and fast-
paced action, Stimwave

19. Synthwave ｜合成風潮

結合 80 年代的合成音樂和現代元素，強調復古和未來感。
a neon-lit cityscape with retro-futuristic elements and
synthesizers, Synthwave

20. Tenwave ｜十風潮

結合 20 世紀 10 年代的美學和電子音樂元素，強調獨特和
現代感。
a cityscape blending elements from the 2010s with
futuristic enhancements, Tenwave

21. Terrorwave ｜恐怖風潮

結合恐怖元素和電子音樂，強調陰森和驚悚的視覺效果。
a dark and eerie cityscape with horror motifs and neon lights, Terrorwave

22. Trillwave ｜高音風潮

結合高音電子音樂元素，強調清脆和高亢的視覺效果。
a vibrant cityscape with high-pitched sounds and neon decorations, Trillwave

23. Tumblewave ｜翻滾風潮

結合動態運動和電子音樂元素，整體強調動感和活力的視覺效果。
acrobats performing in a neon-lit urban environment, Tumblewave

24. Vaporwave ｜蒸汽波

結合 80 年代和 90 年代的美學元素，強調懷舊和夢幻的視覺效果。
a surreal landscape with retro-futuristic elements and pastel colors, Vaporwave

05. 學院風格（Academia Styles）💻 391

01. Adventure Pulp｜冒險小說

結合了冒險和學術元素，強調探索和英雄主義。
an explorer discovering ancient ruins in a dense jungle,
Adventure Pulp

02. Art Academia｜藝術學院

強調藝術創作和學術研究，結合創意和學術氛圍。
students painting and sculpting in a classic art studio,
Art Academia

03. Ballet Academia｜芭蕾學院

結合芭蕾舞和學術元素，強調優雅和紀律。
ballerinas practicing in a grand rehearsal hall, Ballet
Academia

04. Bibliopunk｜書籍龐克

結合書籍和龐克元素，強調反叛和學術的結合。
students reading punk literature in a grungy library,
Bibliopunk

05. Chaotic Academia │混亂學院

強調學術研究中的混亂和創新，結合不拘一格的風格。
a chaotic classroom with students engaged in various
creative projects, Chaotic Academia

06. Classic Academia │經典學院

強調傳統學術氛圍和經典教育，結合古典和嚴謹的風格。
students studying ancient texts in a grand, old library,
Classic Academia

07. Cryptid Academia │神祕生物學院

結合神祕生物和學術研究，強調神祕和探險。
researchers documenting sightings of cryptids in a
dense forest, Cryptid Academia

08. Dark Academia │黑暗學院

結合哥特風和學術元素，強調黑暗和神祕的氛圍。
students studying by candlelight in a dimly lit, gothic
library, Dark Academia

09. Dark Paradise ｜黑暗天堂

結合黑暗和美麗的景象，強調矛盾和吸引力。
a serene yet eerie garden under a dark sky, Dark
Paradise

10. Darkest Academia ｜最黑暗學院

結合極端黑暗元素和學術研究，強調恐怖和神祕。
scholars delving into forbidden texts in a shadowy,
ancient library, Darkest Academia

11. English Major ｜英語專業

強調英語文學和語言研究，結合學術和文化元素。
students discussing classic literature in a cozy, book-
lined room, English Major

12. Fairy Academia ｜精靈學院

結合精靈和學術元素，強調奇幻和神祕的氛圍。
scholars studying magical creatures in an enchanted
forest, Fairy Academia

13. Goblin Academia ｜哥布林學院

結合哥布林和學術元素，強調奇異和幽默的氛圍。
researchers observing goblins in a fantastical laboratory,
Goblin Academia

14. Green Academia ｜綠色學院

強調環保和自然研究，結合學術和生態元素。
students conducting experiments in a lush, green
botanical garden, Green Academia

15. Grey Academia ｜灰色學院

結合中性色調和學術氛圍，強調冷靜和嚴謹。
students studying in a minimalist, grey-toned library,
Grey Academia

16. Horror Academia ｜恐怖學院

結合恐怖元素和學術研究，強調驚悚和神祕的氛圍。
scholars exploring haunted libraries and writing about
supernatural phenomena, Horror Academia

17. Internet Academia ｜網絡學院

結合現代科技和學術研究，強調數字化和連接性。
students engaging in online classes and digital research in a high-tech environment, Internet Academia

18. Light Academia ｜輕學院

強調積極和明亮的學術氛圍，結合歡樂和知識。
students studying in a sunlit library with open windows and fresh air, Light Academia

19. Lit Kid ｜文學小孩

強調青少年文學和創作，結合天真和創意。
young students writing stories and reading in a vibrant classroom, Lit Kid

20. Miscellaneous Academia ｜各種學院

包含多樣學術主題和研究，強調多樣性和廣泛的知識。
a diverse group of students studying various subjects in a lively environment, Miscellaneous Academia

21. Musical Academia │ 音樂學院

結合音樂和學術元素，強調創意和聲音的氛圍。
students composing and playing instruments in a grand music hall, Musical Academia

22. Ocean Academia │ 海洋學院

結合海洋研究和學術元素，強調探索和自然的視覺效果。
marine biologists studying sea life in an underwater lab, Ocean Academia

23. Pastel Academia │ 粉彩學院

結合柔和色調和學術氛圍，強調溫暖和舒適的環境。
students studying in a cozy room with pastel-colored walls and furniture, Pastel Academia

24. Progressive Academia │ 進步學院

強調前沿研究和創新思維，結合現代和未來元素。
researchers using cutting-edge technology in a futuristic lab, Progressive Academia

25. Queer Academia ｜酷兒學院

結合 LGBTQ+ 文化和學術研究，強調包容和多樣性。
students discussing queer theory and gender studies in an inclusive classroom, Queer Academia

26. Robotics Kids ｜機器人小孩

結合兒童和機器人技術，強調創新和趣味性。
children building and programming robots in a modern lab, Robotics Kids

27. Romantic Academia ｜浪漫學院

結合浪漫主義和學術元素，強調情感和美學。
students reading poetry and writing love letters in a picturesque garden, Romantic Academia

28. Schoolgirl Lifestyle ｜女學生生活

結合女學生的日常生活和學術氛圍，強調青春和活力。
schoolgirls in uniforms studying and chatting in a lively classroom, Schoolgirl Lifestyle

29. Science Academia ｜科學學院

結合科學研究和學術元素，強調探究和實驗。

scientists conducting experiments and analyzing data in a high-tech lab, Science Academia

30. Studyblr ｜學習博客

結合博客文化和學術氛圍，強調分享和互動。

students posting study tips and notes on their blogs in a modern study space, Studyblr

31. Studyplace ｜學習場所

強調舒適和高效的學習環境，結合現代和傳統元素。

a cozy and well-equipped study room with books and digital devices, Studyplace

32. Theatre Academia ｜戲劇學院

結合戲劇藝術和學術研究，強調表演和創作。

actors rehearsing and performing in a grand theater, Theatre Academia

33. Vibrant Academia ｜活力學院

強調積極和充滿活力的學術氛圍，結合色彩和能量。
students engaging in dynamic discussions and activities in a colorful environment, Vibrant Academia

34. Vintage Academia ｜復古學院

結合復古風格和學術氛圍，強調懷舊和經典。
students using vintage typewriters and reading old books in a classic library, Vintage Academia

35. Witchy Academia ｜巫師學院

結合巫術和學術元素，強調神祕和奇幻。
students studying ancient spells and brewing potions in a magical classroom, Witchy Academia

36. Writer Academia ｜作家學院

結合寫作和學術研究，強調創作和文學。
writers drafting novels and sharing ideas in a serene, book-filled room, Writer Academia

01. Anglo Goth ｜英式哥德

結合英國傳統和哥德元素，強調古典和陰鬱的氛圍。
An image in the style of Anglo Goth

02. Bubble Goth ｜泡泡哥德

結合哥德和泡泡糖流行元素，強調色彩繽紛和反差的效果。
An image in the style of Bubble Goth

03. Cybergoth ｜賽博哥德

結合未來科技和哥德元素，強調霓虹燈和科技感。
An image in the style of Cybergoth

04. Goth ｜哥德

經典哥德風格，強調黑暗、神祕和浪漫的元素。
An image in the style of Goth

05. Health Goth ｜健康哥德

結合健身和哥德元素，強調運動和健康生活方式。
An image in the style of Health Goth

06. Mallgoth ｜商場哥德

結合哥德和青少年流行文化，強調日常和反叛精神。
An image in the style of Mallgoth

07. Midwest Goth ｜中西部哥德

結合美國中西部文化和哥德元素，強調荒涼和神祕感。
An image in the style of Midwest Goth

08. Nu-Goth ｜新哥德

結合現代和哥德元素，強調簡約和時尚感。
An image in the style of Nu-Goth

09. Pastel Goth ｜粉彩哥德

結合哥德和粉彩元素，強調反差和柔和色彩。
An image in the style of Pastel Goth

10. Regional Goth ｜地區哥德

結合特定地區文化和哥德元素，強調地方特色和陰鬱氛圍。
An image in the style of Regional Goth

11. Romantic Goth ｜浪漫哥德

結合浪漫主義和哥德元素，強調情感和美學。
An image in the style of Romantic Goth

12. Southern Goth ｜南方哥德

結合美國南方文化和哥德元素，強調歷史和神祕感。
An image in the style of Southern Goth

13. Suburban Goth ｜郊區哥德

結合郊區生活和哥德元素，強調日常和陰鬱的視覺效果。
An image in the style of Suburban Goth

14. Trad Goth ｜傳統哥德

經典哥德風格，強調黑暗、美學和復古元素。
An image in the style of Trad Goth

15. Victorian Goth ｜維多利亞哥德

結合維多利亞時代和哥德元素，強調古典和華麗的氛圍。
An image in the style of Victorian Goth

16. Woodland goth ｜森林哥德

結合自然和哥德元素，強調森林和神祕感。
An image in the style of Woodland goth

01. Amekaji ｜美式休閒風

A person dressed in casual American-inspired streetwear, Amekaji

02. Angura Kei ｜地下風

A solitary figure in dark, avant-garde attire with traditional elements, Angura Kei

03. Bōsōzoku ｜暴走族

A rebellious individual in biker gang-inspired clothing, Bōsōzoku

04. Cult Party Kei ｜宗教派對風

A whimsical individual dressed in vintage-inspired, ethereal outfit, Cult Party Kei

05. Decora Kei │ 裝飾風

A colorful person adorned with bright, decorative accessories, Decora Kei

06. Dolly Kei │ 洋娃娃風

A single person in a vintage, fairytale-inspired doll-like outfit, Dolly Kei

07. Fairy Kei │ 童話風

A pastel-clad person in a cute, fantasy-inspired outfit, Fairy Kei

08. Ganguro │ 顏黑風

A tanned individual with bleached hair and colorful makeup, Ganguro

09. Girly Kei | 少女風

A person in sweet, feminine clothing with soft colors, Girly Kei

10. Gyaru | 辣妹風

A trendy girl in glamorous outfit with heavy makeup and big hairstyle, Gyaru

11. Jirai Kei | 地雷系

A person in dark, edgy fashion with contrasting soft elements, Jirai Kei

12. Kimono Style | 和服風

A person dressed in traditional Japanese kimono with modern accessories, Kimono Style

13. Kogal │女高中生風

A high school girl in a modified uniform with loose socks and trendy accessories, Kogal

14. Larme Kei │柔媚風

A romantic individual in a soft, feminine, magazine-style outfit, Larme Kei

15. Mori Kei │森女風

A serene figure in natural, layered attire inspired by nature, Mori Kei

16. Lolita │蘿莉塔風

A single person in an elaborate, Victorian-inspired dress with lace and frills, Lolita

17. Mote Kei ｜受歡迎風

A trendy person in mainstream, attractive fashion, Mote Kei

18. Onii Kei ｜兄貴風

A cool young man in stylish, flashy attire, Onii Kei

19. Oshare Kei ｜時尚風

A fashionable individual in bright, stylish clothing, Oshare Kei

20. Party Kei ｜派對風

A lively person in fun, eye-catching party attire, Party Kei

21. Peeps │美式學院風

A person dressed in American preppy style clothing, Peeps

22. Pop Kei │流行風

A vibrant figure in colorful, playful outfit inspired by pop culture, Pop Kei

23. Salon Kei │沙龍風

A sophisticated person in elegant, salon-style attire, Salon Kei

24. Tanbi Kei │耽美風

A beautifully dressed individual in elaborate, fantasy-inspired outfit, Tanbi Kei

25. Visual Kei ｜ 視覺系

A striking figure in dramatic, rock-inspired outfit and makeup, Visual Kei

08. 時代風格（TimePeriod） 💻 410

01. Prehistoric Art ｜ 史前藝術

A cave painting of a beautiful woman, Prehistoric Art style

02. Ancient Art ｜ 古代藝術

A painting of a beautiful woman from an ancient civilization, Ancient Art style

03. 500–1000 CE ｜ 公元 500–1000 年

A medieval painting of a beautiful woman in traditional attire, 500–1000 CE style

04. 1000–1400 CE ｜ 公元 1000–1400 年

A medieval painting of a beautiful woman in traditional attire, 1000–1400 CE style

05. 15th Century ｜ 15 世紀

A Renaissance painting of a beautiful noblewoman, 15th Century style

06. 16th Century ｜ 16 世紀

A painting of a beautiful woman in elaborate court attire, 16th Century style

07. 17th Century | 17 世紀

A Baroque painting of a beautiful woman at a royal banquet, 17th Century style

08. 18th Century | 18 世紀

A Rococo painting of a beautiful woman in a lavish garden, 18th Century style

09. 19th Century | 19 世紀

A Victorian-era painting of a beautiful woman in period attire, 19th Century style

10. Late 19th Century | 19 世紀末

An Impressionist painting of a beautiful woman at a Parisian café, Late 19th Century style

11. 1860–1969 | 1860–1969 年

A historical painting of a beautiful woman transitioning from 19th to 20th century, 1860–1969 style

12. 1900–1917 | 1900–1917 年

An Art Nouveau painting of a fashionable beautiful woman, 1900–1917 style

13. 1918–1939 (Interwar) | 1918–1939 年（戰間期）

An Art Deco painting of a beautiful woman in a Jazz Age nightclub, 1918–1939 (Interwar) style

14. 1940s–1950s | 1940 年代 –1950 年代

A retro painting of a beautiful woman in a vintage diner, 1940s–1950s style

15. 1960s │ 1960 年代

A pop art painting of a beautiful woman at a music festival, 1960s style

16. 1970s │ 1970 年代

A disco-themed painting of a beautiful woman dancing, 1970s style

17. 1980s │ 1980 年代

A neon-lit painting of a beautiful woman in bold, colorful attire, 1980s style

18. 1990s │ 1990 年代

A grunge-inspired painting of a beautiful woman at a skate park, 1990s style

19. 2000s │ 2000 年代

A modern painting of a beautiful woman in contemporary fashion, 2000s style

20. 2010s │ 2010 年代

A stylish painting of a beautiful woman in trendy attire, 2010s style

21. 2020s │ 2020 年代

A futuristic painting of a beautiful woman in avant-garde fashion, 2020s style

女神 1

女神 2

PART 5

AI 影像生成，
人文素養是關鍵義

透過前面四個章節的介紹，主要是要讓大家更了解 AI 影像生成的邏輯，邏輯通了，你會比一般人更快掌握 AI 生成要訣。也整理了各種「風格提示詞」方便學習使用。但更重要的是利用 AI 生成式影像工具進行創作時，人文素養的發揮才是關鍵，豐富的人文素養方足以提升作品的深度和質量，也讓創作更具意義和文化價值。本書所強調的人文素養，包括：

1. 結合歷史背景 🖥 418

在進行影像創作時，運用歷史知識可以賦予作品更多的背景和深度。

生成範例｜ 18 世紀法國古典主義建築宮殿

當你想生成一幅古典主義風格的建築圖像時，對該時代的建築特徵、社會背景和文化風貌的了解，能幫助你撰寫更精確的提示詞，讓 AI 生成更符合歷史真實性和美學標準的作品。

An 18th-century French classical architectural style palace with towering columns and symmetrical facades, set against the backdrop of the royal gardens of the time. --ar 3:2
18 世紀法國古典主義建築風格的宮殿，擁有高聳的圓柱和對稱的立面，背景是當時的皇家園林。

將上圖生成後的影像進行後製加工,增加一些書法及國畫元素,一張更完整的作品便誕生了!

生成範例 │ 中國觀音佛祖水墨畫

以佛教人物結合中國地理作為題材,配合中國水墨畫表現創作一幅畫面。

A full-body portrait of Guanyin Budda, with the landscape of Zhangjiajie Mountains in China as the background. Among the mountain ranges, there is a dragon and a phoenix flying. The artwork is in the style of traditional Chinese black and white ink painting, with no color. The background features ink wash effects. The figure of Guanyin is depicted with simple brush lines. --v 6.0 --ar 2:3

觀音佛祖全身像,以中國張家界山風景為背景。山脈之中,有龍鳳飛翔。作品採用中國傳統黑白水墨畫風格,無色彩。背景具有水墨效果。觀音像以簡單的筆觸描繪

2. 強化生活觀察

對生活細節的觀察能使 AI 生成的影像更生動且具體。當你要創作一幅日常生活場景的圖像時，將觀察到的細節融入提示詞中，能讓影像更貼近現實，富有情感和故事性。

生成範例 | 黃昏下的看書老人

At dusk, an old man was reading a book on a park bench. The breeze blew gently, the leaves rustled, and the afterglow of the setting sun shone on the ground through the leaves. --ar 3:2

黃昏時分，一位老人在公園長椅上看書，微風輕拂，樹葉沙沙作響，夕陽的餘暉透過樹葉灑在地上。

生成範例 | 創作一張自畫像

A cinematic, behind-the-scenes photograph of an Asian bald man with glasses working as a film director on a movie set. He is holding a megaphone in one hand and gesturing towards the scene being filmed with the other. The background shows a bustling film set with crew members, cameras, and lighting equipment. The man's expression is focused and authoritative as he directs the action, captured in the dynamic environment of filmmaking. The lighting highlights the intensity and creativity of the film production process. --ar 3:2

一張電影幕後照片，拍攝的是一名戴眼鏡的亞洲光頭男子在電影片場擔任電影導演。他一手拿著擴音器，另一根手指向正在拍攝的場景。背景是一個熙熙攘攘的電影佈景，有工作人員、攝影機和照明設備。在電影製作的動態環境中捕捉到的動作時，該男子的表情專注而權威。燈光凸顯了電影製作過程的強度和創造力。

發現了嗎？這樣的描述方式是不可能生成自己的臉孔面貌，這裡提供一個有趣的換臉小工具（https://tw.vidnoz.com），操作上非常簡單，只要先準備幾張欲置換的臉孔，然後根據網站上的提示進行即可。

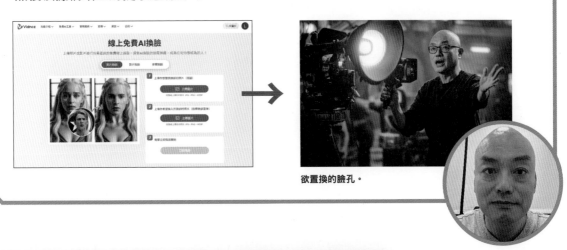

欲置換的臉孔。

當然不換臉也行，我們可以避開真實照片的效果，改以繪畫方式表現。

Joseph Zbukvic's watercolor painting depicting of A bald Asian 30-years-old man with thick, upturned eyebrows, double eyelids, eye bags, and furrowed brows, wearing black-framed glasses, sits alone on a balcony at night, holding a glass of red wine and smoking a cigarette, gazing at the Taipei streets outside. --style raw --v 6.0 --ar 3:2

以約瑟夫‧茲布克維奇（Joseph Zbukvic）的水彩畫風格描繪了一個三十多歲的光頭亞洲男子，眉毛濃密，上翹，雙眼皮，有眼袋，眉頭緊鎖，戴著黑框眼鏡，晚上獨自坐在陽台上，手裡拿著一杯紅酒喝著酒，抽著煙，望著外面的台北街道。

3. 融合藝術修養 📺 422

理解和運用不同的藝術風格，可以在 AI 創作中實現風格的多樣性和美學的高度。例如，當你想要創作一幅印象派風格的圖像時，熟悉印象派的色彩運用和筆觸特徵能讓你更好地引導 AI 生成符合該風格的作品。

Impressionist painting in the style of Monet, a field of blooming flowers, soft light and shadows and airy colors. --ar 4:3

莫內風格的印象派畫作，一片繁花盛開的田野，柔和的光影和輕快的色彩。

利用同樣表現方式的提示詞，修改其中的一些關鍵字。例如這個範例，我們僅需要修改藝術家（達文西、維梅爾）及作品（蒙娜麗莎、戴珍珠耳環的少女、抱銀貂的女子）名稱，就可以延伸出更多不同且有趣的創。

生成範例｜藝術經典延伸創作

生成範例｜超現實派《戴珍珠耳環的少女》

我們再以《抱銀貂的女子》（Lady with an Ermine）進行創作。我們也使用 midjouney 的「人物參照」參數 --cref (url= 參照影像的網址)，提供原始畫作影像作為生成的依據，並設定 –cw 100 讓結果保留最大的影像參照值。

生成範例｜超現實派《抱銀貂的女子》

A photorealistic scene of the Lady with an Ermine painting where the portrait transforms into a real person, emerging from the painting with the original background and frame intact. The woman is dressed in Renaissance attire, perfectly matching the iconic painting, and she holds an ermine in her arms. She is depicted actively stepping out of the frame, with her upper body and one hand already outside, creating a dynamic and interactive composition. Her face retains the serene expression characteristic of the Lady with an Ermine. The original background of the painting is preserved, enhancing the surreal and imaginative atmosphere. The ornate frame around the painting remains, emphasizing the transition from art to reality. The lighting captures the essence of da Vinci's style, with soft highlights and gentle shadows. --style raw --v 6.0 --cref https://s.mj.run/5RTylnw3KA --cw 100 --ar 2:3

抱銀貂的女子 (Lady with an Ermine) 的逼真場景，肖像變成真人，從畫中浮現出來，原始背景和框架完好無損。這位女士穿著文藝復興時期的服裝，與這幅標誌性的畫作完美搭配，她懷裡抱著一隻貂。她被描繪成積極地走出畫面，上半身和一隻手已經在外面，創造出動態和互動的構圖。她的臉上保留著銀貂女士特有的平靜表情。畫面保留了原有的背景，增強了超現實和想像的氛圍。畫作周圍的華麗框架仍然存在，強調從藝術到現實的過渡。燈光捕捉了達文西風格的精髓，柔和的高光和柔和的陰影。

4. 運用文學素養 💻 424

目前的主流地 AI 生成方式仍是以文生圖（text to image）為主，因此如果平素有閱讀習慣，且累積大量文本材料，例如各類中西方文學名著，像是金庸小說、哈利波特、紅樓夢、西遊記等等，這些都是可以大量獲得靈感的來源。當然我們也可以自行創作，總之，文學素養幫助你以更生動、具象的方式撰寫提示詞，使生成的影像充滿情感和氛圍。

生成範例｜以文生圖

A quiet lakeside, filled with mist in the morning, the boat swaying gently in the distance, and the surrounding trees are poetically reflected in the morning light. --ar 4:3

一個寧靜的湖邊，清晨的霧氣彌漫，遠處的小船輕輕搖晃，周圍的樹木在晨曦中如詩般映襯。

透過結合歷史背景、生活觀察、藝術修養、文學素養，甚至哲學思考，創作者可以利用 AI 生成式影像工具創作出更有深度和文化內涵的作品。目前為止 AI 影像生成，甚至不只影像生成，包括動畫、音樂等等，透過自然語言模型，都是一種文字的轉換能力，而本書想傳達的一個重點，與其說是影像生成，不如說是「說故事」，而我們要學習的就是說故事的能力，這能力包括知識含量、邏輯推理及各種元素的交叉組成。

有時我們甚至可以直接利用一些文本進行圖像創作，例如《紅樓夢》，我們以書中腳色王熙鳳為例，曹雪芹是這樣描寫王熙鳳的：「……彩繡輝煌，恍若神妃仙子：頭上戴著金絲八寶攢珠髻，綰著朝陽五鳳桂珠釵；項上戴著赤金盤螭瓔珞圈；裙邊繫著豆綠宮縧，雙衡比目玫瑰佩；身上穿著縷金百蝶穿花大紅洋緞窄褃襖，外罩五彩刻絲石青銀鼠褂；下著翡翠撒花洋縐裙。一雙丹鳳三角眼，兩彎柳葉吊梢眉；身量苗條，體態風騷；粉面含春威不露，丹唇未啟笑先聞。」。我們將這段話用改寫成白話文，再加上一些畫面元素，變成提示詞生成畫面如下：

生成範例｜王熙鳳

A beautiful woman, aged 30, in the Ming Dynasty of China. She has a slender figure, exuding elegance; with A pair of slender eyebrows, and slim eyes, eye shape in which the outer corners of the eyes are turned up. Adorning her head is a hairpin embellished with eight treasures and pearls, entwined with a hairpin of five phoenixes and laurel beads; around her neck hangs a pendant necklace of red-gold dragon; her skirt hem is tied with a bean-green palace sash, adorned with rose pendants; she wears a narrow crimson satin jacket embroidered with gold butterflies and flowers, layered with a multicolored silk-lined slate-blue fur coat; and she is dressed in an emerald pleated skirt. It's in the style of a cinematic shot with high resolution and high definition in very detailed --style raw --v 6.0 --ar 1:2

畫面中的書法非 AI 生成，是在網上下載書法作品後，再透過 photoshop 合成。

我們也可嘗試自行設計動畫或遊戲腳色，以金庸作品中的人物小龍女、黃蓉、黃藥師、歐陽鋒、段智興、洪七公等為例，根據小說描述他們的外貌特徵、服飾、武學招式或武器。

1. 生成範例｜白衣勝雪的小龍女

A cute martial arts cartoon character, a 15-year-old girl from the Song Dynasty of China. She is beautiful but aloof, with willow leaf eyebrows, large eyes, and wears a long white dress with gloves. Her complexion is pale. Her weapon is a white ribbon with a golden ball at the end. In the style of 3D Pixar cartoon film. --ar 3:4 --niji 6 --s 750 --style raw

2. 生成範例｜古靈精怪的黃蓉

A cute version of a martial arts cartoon character. Song Dynasty, China. Female, 15 years old, a beautiful and playful girl with willow-leaf eyebrows and big eyes. She wears clothes resembling those of a beggar, with a slightly dirty face. Her weapon is a flute. In the style of 3D Pixar cartoon film. --ar 3:4 --niji 6 --s 750 --style raw

3. 生成範例｜東邪黃藥師

A cute martial arts cartoon character from the Song Dynasty of China. He is a 50-year-old man with a tall and thin stature, sporting a goatee. He wears a green robe and has a stern expression. His weapon is a green flute. In the style of 3D Pixar cartoon film. --ar 3:4 --niji 6 --s 750 --style raw

4. 生成範例｜西毒歐陽鋒

A cute martial arts cartoon character from the Song Dynasty of China. He is a 50-year-old man with a stout build, sporting a curled beard. He wears a grayish-white robe and traditional Chinese gold ornaments on his head. He has a fierce expression. His weapon is a snake staff with two poisonous snakes, one black and one white, on the head of the staff. In the style of 3D Pixar cartoon film. --ar 3:4 --niji 6 --s 750 --style raw

5. 生成範例｜南帝段智興

A cute martial arts cartoon character from the Song Dynasty of China. He is a 40-year-old man with a tall and thin stature, sporting a goatee. He wears the imperial dragon golden robe and crown, with meditative posture. His martial arts skill is the Finger pointing technique. In the style of 3D Pixar cartoon film. --ar 3:4 --niji 6 --s 750 --style raw

6. 生成範例｜北丐洪七公

A cute martial arts cartoon character from the Song Dynasty of China. He is a 40-year-old beggar male with a stout stature, messy hair, and beard. His clothes are tattered with patches, and he has nine small cloth bags hanging from his waist. He has a gentle expression and walks barefoot. Hi s weapon is a green bamboo stick. In the style of 3D Pixar cartoon film. --niji 6 --s 750 --style raw --ar 3:4

如之前所提及，目前這些影像生成的模型訓練，普遍缺乏東方的語料，當然 AI 就不會懂得金庸小說中的這些人物，所以我們就必須鉅細靡遺的針對人物特徵進行描寫，這也是培養觀察的能力。但如果是要創作西方人物，那就真的方便多了，例如：哈利波特小說中的人物，只需要加入媒材及風格描寫，人物主題就只需要人名。

生成範例｜哈利波特

Joseph Zbukvic's watercolor painting depicting a full body portrait of Harry Potter --style raw --s 750 --ar 3:4 --v 6.0

生成範例｜阿不思・鄧不利多

Joseph Zbukvic's watercolor painting depicting a full body portrait of Albus Dumbledore --style raw --s 750 --ar 3:4 --v 6.0

但這樣的創作就顯得無趣，接下來的 AI 創作都屬於影像模型較缺乏的東方甚或台灣題材，也是希望透過我小小的力量，能讓 AI 模型接觸到屬於我們的文化。

我們不斷強調 AI 影像生成就是將主題清晰的描述，配合指定的媒材、風格，例如：

生成範例｜外星人在西藏

Real photography, in front of the huge thangka mural in Tibet, with an image of an alien in the center of the mural. Science fiction, future world, cyberpunk, high resolution, high definition, rich in details, bright colors, very much in the style of National Geographic photography.

真實攝影，在西藏巨大的唐卡壁畫前，壁畫中央有一個外星人的圖像。科幻，未來世界，賽博朋克，高解析度，高清，細節豐富，色彩鮮豔，很有國家地理攝影的風格。

A painting in the style of Lorenzo Lotto, depicting Siddhartha Gautama (Shakyamuni Buddha) achieving enlightenment under the Bodhi tree. The scene features the Buddha seated in a serene meditative posture beneath the expansive branches of the Bodhi tree. His expression is one of profound peace and enlightenment. The background showcases the natural beauty of the surroundings, with lush foliage, blooming flowers, and a soft, golden light that bathes the scene, reminiscent of Lorenzo Lotto's use of vibrant colors and detailed backgrounds. The setting includes subtle details of nature, such as birds, small animals, and flowing streams, enhancing the tranquility of the moment. The overall style incorporates Lotto's characteristic attention to detail, rich color palette, and a harmonious composition that emphasizes the spiritual significance of the Buddha's enlightenment. The painting captures the moment of awakening with a sense of reverence and timeless beauty. --ar 3:2 --style raw --v 6.0

An old woman wearing reading glasses was mending a piece of clothing and making threading movements. The background is an old house from the 1950s in Taiwan. The decoration is very simple. The sun shines in through the window, and the figures are backlit, by Mark Shaw --ar 2:3 --style raw --v 6.0

一位戴著老花眼鏡的老婦人正在縫補一件衣服，並做著穿線的動作。背景是台灣一棟 20 世紀 50 年代的老房子。裝修很簡單。陽光透過窗戶照進來，人物是背光的，馬克・肖拍攝

429

當然也可以生成一張來自未來世界
的科技送子觀音。

生成範例｜未來世界的送子觀音

A photorealistic image depicting a robotic version of the Goddess of Mercy（送子觀音）in a cyberpunk style, set against a clean, futuristic background. The robotic Guanyin is designed with sleek, metallic components, glowing accents, and intricate cybernetic details. Her serene and gentle expression is maintained, with LED lights forming her eyes and a holographic aura surrounding her. She holds a child, also depicted with robotic features, symbolizing the blessing of fertility and compassion in a futuristic context. The child's design mirrors the advanced technology and cybernetic enhancements of Guanyin. The background is minimalistic and clean, featuring a high-tech cityscape with sleek skyscrapers, neon lights, and floating holograms. The setting emphasizes a futuristic and sci-fi aesthetic, with elements of advanced technology seamlessly integrated into the environment. The overall style emphasizes the spiritual significance and maternal compassion of Guanyin, blending the characteristic elements of cyberpunk and futuristic design to create a visually stunning and imaginative image. The interplay of light and shadow, along with the use of neon colors and metallic textures, adds depth and a sense of realism to the scene. --ar 2:3 --style raw --v 6.0

▶ AI 創作大賞析

AI 影像生成工具是人所設計出來的，他不是魔法，自然不會有咒語，但如何精確地將心中所想表達出來，則需要更多的敘事能力。在最後這個章節，我們根據以上五個章節內容，提供了許多創作範例，包括：繪畫創作、攝影創作、產品設計、室內設計、動漫人物、繪本創作等。

此外也示範如何利用 Midjourney 進行一些主題系列性的創作，例如：科幻超現實、世界采風、山道猴子、十二生肖等。透過這些不同主題的創作，讀者可以學習到更多的提示詞技巧，除了進行一些創意圖像外，也可以適當地具體應用在一些工作上，例如：製作室內設計圖與設計師溝通裝修風格，建立自己所要的產品設計概念圖與設計師溝通等等。

繪畫創作 LOOK 🖥 431

01. 台北傳統菜市場

Oil painting depicting a bustling traditional market in Taipei on a summer morning. The scene captures vendors arranging fresh fruits and vegetables, early morning sunlight creating dappled shadows on the cobblestone paths, with locals shopping and conversing. In the style of Impressionism by Claude Monet. --ar 3:2

02. 九份老街

Oil painting depicting the bustling streets of Jiufen at dusk. The scene captures red lanterns glowing along narrow alleyways, with people strolling and exploring shops. The distant mountains and ocean are visible under a fading sky. In the style of Post-Impressionism. --ar 3:2

03. 淡水漁港

Watercolor painting capturing the tranquil atmosphere of Tamsui Fisherman's Wharf at sunrise. The scene shows fishing boats gently bobbing in the calm water, with the sun casting a soft pink and orange hue over the horizon. In the style of J.M.W. Turner. --ar 3:2

04. 台東稻田

Oil painting showcasing the vast rice fields of Taitung in the golden hour. The scene features farmers working in the fields, with the golden light reflecting off the ripening crops and surrounding mountains. In the style of Realism. --ar 3:2

05. 龍山寺

Traditional ink and wash painting depicting the iconic Longshan Temple in Taipei during a rainy evening. The scene includes worshippers with umbrellas, incense smoke rising, and the temple's intricate architecture. In the style of Chinese traditional painting. --ar 3:2

06. 玉山積雪

Oil painting showcasing the snow-capped peaks of Yushan under a clear winter sky. The scene features pristine snow covering the mountains and pine trees, with a bright blue sky overhead. In the style of Romanticism. --ar 3:2

07. 士林夜市

Oil painting depicting the vibrant energy of Shilin Night Market in Taipei. The scene captures bustling food stalls, colorful lights, and crowds of people enjoying street food. In the style of Fauvism. --ar 3:2

08. 太魯閣峽谷

Pastel drawing capturing the dramatic cliffs and lush greenery of Taroko Gorge. The scene includes the Liwu River winding through the gorge, with mist rising from the water. In the style of Caspar David Friedrich. --ar 3:2

09. 日月潭

Watercolor painting depicting the serene beauty of Sun Moon Lake at dawn. The scene shows the lake's calm waters reflecting the surrounding mountains and a lone boat gently gliding across the surface. In the style of Traditional Chinese landscape painting. --ar 3:2

10. 基隆港

Oil painting capturing the busy activity at Keelung Harbor during a summer afternoon. The scene features cargo ships, cranes, and workers loading and unloading goods, with the harbor bathed in warm sunlight. In the style of Social Realism. --ar 3:2

11. 北投溫泉

Watercolor painting depicting the tranquil hot springs of Beitou, surrounded by lush greenery and steam rising from the waters. The scene includes traditional wooden bathhouses and stone paths leading through the forest. In the style of Japanese woodblock prints. --ar 3:2

12. 嘉義阿里山鐵路

Oil painting showcasing the historic Alishan Forest Railway in Chiayi, winding through dense forests and steep mountainsides. The scene captures the iconic red train making its way through the misty morning landscape. In the style of Realism. --ar 3:2

13. 澎湖群島

Pastel drawing capturing the serene beaches and traditional stone houses of the Penghu Islands. The scene features turquoise waters, coral reefs, and fishing boats anchored near the shore under a clear blue sky. In the style of Impressionism. --ar 3:2

14. 台南古城

Oil painting depicting the historic architecture and streets of Tainan's old town. The scene captures traditional buildings with red brick walls and tiled roofs, with locals going about their daily lives. In the style of Post-Impressionism. --ar 3:2

15. 恆春古城

Watercolor painting showcasing the ancient walls and gates of Hengchun Old Town, surrounded by lush tropical vegetation. The scene captures the historic architecture and the calm, warm atmosphere of southern Taiwan. In the style of Romanticism. --ar 3:2

16. 阿里山日出

Ukiyo-e showcasing the breathtaking sunrise over Alishan. The composition features the iconic sea of clouds with the sun rising above the peaks, casting a warm, golden glow over the forested landscape. The distant mountains are softly outlined, creating a peaceful atmosphere. In the style of Katsushika Hokusai. --ar 3:2

17. 台南孔廟

Traditional ink painting capturing the Tainan Confucius Temple in the early morning. The scene includes ancient trees with sprawling roots, traditional red-pillared architecture, and the quiet courtyard filled with soft morning light. In the style of Chinese literati. --ar 3:2

18. 淡水老街

Oil painting depicting the vibrant activity of Tamsui Old Street at sunset. The scene includes bustling shops, street vendors, and people enjoying the view of the river as the sky turns shades of orange and pink. In the style of Fauvism. --ar 3:2

19. 花蓮七星潭

Watercolor painting capturing the rocky shores and clear blue waters of Qixingtan Beach in Hualien. The scene features pebbles on the shore, gentle waves, and distant mountains under a bright, clear sky. In the style of Japanese landscape painting. --ar 3:2

20. 台北 101

Oil painting showcasing Taipei 101 towering over the city skyline at dusk. The scene captures the skyscraper illuminated by the warm colors of sunset, with the bustling city below. In the style of Contemporary Realism. --ar 3:2

攝影創作 LOOK

01. 都市夜景與車軌

A long-exposure photograph capturing
the vibrant energy of a city at night.
The scene features a busy intersection
with streaks of light from passing
cars creating dynamic trails. The
composition focuses on the contrast
between the bright city lights and the
dark night sky. The use of a tripod and
a wide-angle lens emphasizes the
depth and movement in the image. --ar
3:2

02. 花朵的微距攝影

A macro photograph focusing on the
delicate details of a blooming flower. The
image captures the intricate patterns
and textures of the petals, with morning
dew drops glistening in the soft natural
light. The shallow depth of field creates a
beautiful bokeh effect in the background,
highlighting the sharp focus on the
flower's center. --ar 3:2

03. 森林中的霧景攝影

A misty forest scene captured in black and
white, emphasizing the mysterious and tranquil
atmosphere. The composition features tall trees
fading into the mist, with soft light filtering through
the branches. The use of a medium format
camera enhances the depth and tonal range of
the image, while the black and white treatment
adds to the ethereal quality of the scene. --ar 3:2

04. 老街上的街頭肖像

A candid street portrait in a busy market street, using a 50mm lens to capture the expression of an elderly vendor selling traditional goods. The natural light creates soft shadows on the subject's face, highlighting the wrinkles and textures that tell a story of years of hard work. The shallow depth of field isolates the subject from the bustling background. --ar 3:2

05. 冰川的動態攝影

A dynamic photograph of a glacier using long exposure to capture the slow movement of the ice. The composition includes the contrast between the icy blue tones of the glacier and the surrounding dark volcanic rock. The use of a neutral density filter allows for a smooth, flowing appearance of the ice, emphasizing the relentless power of nature. --ar 3:2

06. 公園中的小鳥微距攝影

A macro photograph of a small bird perched on a branch in a quiet park. The photograph captures the intricate details of the bird's feathers and the sharpness of its beak and eyes. The background is blurred with a bokeh effect, created by the natural light filtering through the surrounding trees. A 100mm macro lens is used to focus on the delicate features of the bird. --ar 3:2

07. 魚眼鏡頭的城市天際線

A fish-eye lens photograph capturing the curvature of a modern city skyline at dawn. The distorted perspective emphasizes the towering skyscrapers, while the soft morning light creates a gradient of colors in the sky. The composition includes reflections of the buildings on a calm river below, adding to the surreal effect of the image. --ar 3:2

08. 紅外線攝影的荒野風光

An infrared photograph of a barren landscape, where the use of infrared light transforms the vegetation into a ghostly white while the sky takes on a deep, almost black tone. The composition highlights the contrast between the desolate ground and the eerily glowing trees. A wide-angle lens is used to capture the vastness of the scene. --ar 3:2

09. 立可拍的街頭快照

A Polaroid street snapshot capturing the spontaneous energy of a busy pedestrian crossing in a metropolitan area. The image features blurred motion of people walking, with a sharp focus on the street signs and buildings in the background. The instant film's natural grain and vintage color tones add a nostalgic feel to the urban scene. --ar 3:2

10. 海邊的日出剪影

A silhouette photograph taken at sunrise on a beach, capturing the dark outline of a lone figure standing by the water's edge. The bright rising sun creates a stunning contrast between the figure and the glowing horizon, with the gentle waves adding texture to the foreground. The use of a 35mm lens ensures a balanced composition between the subject and the surrounding scenery. --ar 3:2

11. 黑白環境人像攝影

A black and white environmental portrait of an elderly woman sitting in her kitchen, surrounded by pots, pans, and a cluttered table. The composition is a close-up, focusing on the textures and contrast of the scene, with natural light streaming through a window, highlighting the woman's thoughtful expression. --ar 4:5

12. 戶外自然光人像攝影

A portrait of a young man standing in a sunflower field at golden hour. The composition is mid-range, capturing the upper half of the man's body with soft, warm light enhancing the natural tones of his skin. The background of sunflowers is slightly out of focus, creating a serene and peaceful atmosphere. --ar 2:3

13. 長曝光夜間肖像攝影

A long-exposure portrait of a woman standing still on a busy city street at night, with the blurred motion of passing cars and pedestrians creating a dynamic backdrop. The composition focuses on the full body of the woman, sharply in focus, contrasting with the motion blur around her. --ar 3:4

14. 色彩對比人像攝影

A vibrant portrait of a young woman with colorful makeup and bright clothing, set against a stark black background. The composition is a close-up, focusing on her face and upper body, with high contrast making her features and expressions stand out vividly. --ar 1:1

15. 室內低光人像攝影

A low-light portrait of a woman sitting by a window in a dimly lit room, with just a small amount of natural light illuminating her face. The composition is mid-range, focusing on her upper body, with soft shadows creating a moody atmosphere. --ar 3:2

16. 重複曝光藝術人像攝影

A creative double-exposure portrait of a man combined with an overlay of a dense forest. The man's face is blended with the trees and foliage, creating a surreal and dreamlike effect. The composition is mid-range, capturing his head and shoulders as the primary focus. --ar 4:5

17. 街頭快照人像攝影

A candid street portrait of a man laughing with friends at an outdoor café. The composition is mid-range, capturing his upper body and the interaction with the environment, with the background slightly out of focus to highlight the moment. --ar 2:3

18. 微距特寫人像攝影

A macro close-up portrait of an elderly man's eye, showing the intricate details of the iris and wrinkles around it. The composition is extremely close-up, focusing entirely on the eye and the surrounding textures. --ar 1:1

19. 創意光影人像攝影

A portrait of a young woman partially lit by a patterned light source, creating intricate shadows on her face and background. The composition is a close-up, focusing on her face and the play of light and shadow, adding to the mysterious mood. --ar 3:2

20. 雙人剪影攝影

A silhouette photograph of a couple standing on a hilltop at sunset, holding hands with the bright setting sun behind them. The composition is wide-angle, capturing their full bodies along with the expansive landscape surrounding them. --ar 2:3

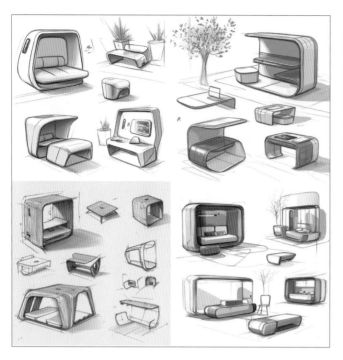

01. 模塊化智能家具

A futuristic modular smart furniture system designed as a hand-drawn sketch, featuring components that can be reconfigured and adapted to different living spaces. The furniture pieces connect via magnetic joints, creating flexible layouts. The design is eco-friendly, with materials that change color based on user preferences. The sketch shows a minimalist approach with clean lines and subtle shading.

02. 懸浮式書桌燈

A conceptual floating desk lamp drawn with markers, showcasing its sleek, minimalist design. The lamp uses magnetic levitation to hover above the desk, with a transparent base to enhance the illusion. The lamp's light source is adjustable, allowing for various color temperatures and brightness levels. The background is a simple gradient color to highlight the product's clean design.

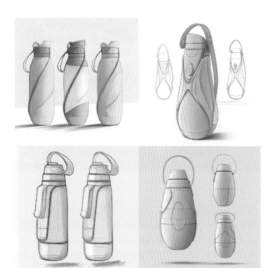

03. 折疊式旅行水壺

A collapsible travel water bottle concept, designed for minimal space usage. The water bottle is made from flexible, durable silicone and folds into a compact, pocket-sized form when not in use. The design includes a built-in filtration system and a loop handle for easy carrying. The sketch is rendered in soft pastels, highlighting its portability and eco-friendly materials.

04. 全息顯示屏手錶

A smartwatch concept with a holographic display, allowing the user to interact with 3D projections directly above the watch face. The design is sleek and futuristic, with a transparent strap and an ultra-thin bezel. The background is a simple, solid color to emphasize the cutting-edge technology. The design is rendered as a digital illustration with sharp, precise lines.

05. 可編程照明系統

A programmable lighting system concept, designed as a hand-drawn sketch. The system consists of modular light panels that can be arranged in various configurations on walls or ceilings. Each panel is touch-sensitive and can be programmed to change colors and brightness based on the user's preferences. The sketch highlights the flexibility and customization of the design with dynamic, flowing lines.

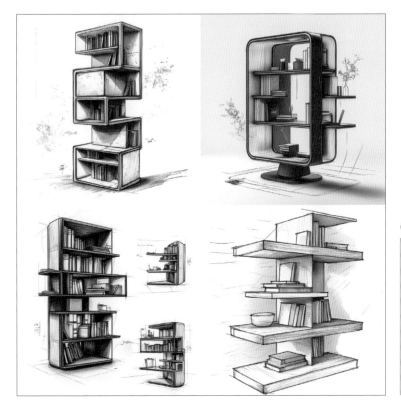

06. 雙面書架

A dual-sided bookshelf concept, designed for small living spaces. The bookshelf has a rotating axis, allowing it to be accessed from both sides, with different shelf configurations on each side. The design is sleek and modern, with a minimalistic frame and a matte finish. The concept is illustrated with a combination of hand-drawn elements and digital shading to showcase the versatility of the design.

07. 應急太陽能充電器

A portable solar-powered charger concept designed for emergency situations. The charger is compact and foldable, with solar panels that can be unfolded to maximize energy absorption. The design includes a rugged, waterproof case and a built-in LED flashlight. The sketch is created with markers, using bright colors to emphasize its utility and durability.

08. 懸浮音響系統

A conceptual floating speaker system that uses magnetic levitation to hover in mid-air. The design is sleek and modern, with the speakers encased in a smooth, spherical shell. The background is a gradient color that enhances the sense of depth and emphasizes the futuristic design. The concept is rendered as a 3D model with realistic textures and lighting.

09. 折疊式屏風

A foldable room divider concept, designed with an emphasis on privacy and aesthetics. The divider features panels made from semi-transparent materials that allow light to pass through while maintaining privacy. The design is minimalistic, with a focus on clean lines and modern textures. The concept is illustrated in a hand-drawn sketch, with subtle shading to show depth and dimension.

10. 便攜式咖啡機

A portable coffee maker concept designed for on-the-go use. The coffee maker is compact and lightweight, with a sleek, cylindrical design. It features a built-in grinder, water reservoir, and heating element, allowing users to brew fresh coffee anywhere. The design is depicted in a clean, minimalist digital illustration, with a focus on its portability and functionality.

11. 懸浮綠植花盆

A floating plant pot concept that uses magnetic levitation to suspend small indoor plants in mid-air. The design is minimalist and modern, with a smooth, ceramic pot that hovers above a wooden base. The background is a simple white to highlight the natural beauty of the plant and the innovative floating mechanism. The concept is depicted as a 3D render with soft lighting and detailed textures.

12. 手提式音響系統

A portable sound system designed with a retro aesthetic. The speakers are encased in a leather-bound box with brass accents, reminiscent of vintage luggage. The concept is illustrated in a detailed hand-drawn sketch, with cross-hatching to show texture and depth. The background features a soft gradient to evoke a nostalgic feel.

13. 智能鏡面

A smart mirror concept that integrates a digital display into a full-length mirror. The mirror shows real-time data such as weather, calendar events, and health stats. The design is sleek and modern, with a frameless mirror and a subtle interface. The concept is rendered as a high-fidelity digital mockup with reflective effects to simulate the mirror's surface.

14. 3D 打印鞋

A pair of concept shoes designed to be 3D printed, with an intricate lattice structure that provides both flexibility and support. The shoes are customizable, allowing users to choose different colors and patterns. The concept is illustrated using a combination of hand-drawn elements and digital rendering, showcasing the unique texture and structure.

15. 交互式書桌

An interactive desk concept designed with a built-in touch-sensitive surface that doubles as a digital workspace. The desk allows users to draw, write, and interact with digital content directly on the surface. The design is modern and minimalistic, with clean lines and a white finish. The concept is presented as a digital illustration with a focus on the seamless integration of technology and furniture.

16. 手繪風格雨具

A conceptual umbrella designed with an artistic, hand-drawn aesthetic. The umbrella features an asymmetrical shape with a canopy that displays a unique hand-drawn pattern, such as abstract shapes or natural elements. The handle is crafted from wood, adding a rustic touch to the modern design. The background is a simple solid color to emphasize the umbrella's artistic quality.

17. 可卷式顯示器

A rollable display concept designed for portable computing. The screen can be rolled up into a compact tube, making it easy to carry and store. The design is ultra-thin and flexible, with a sleek, metallic finish. The concept is illustrated in a digital render with emphasis on the screen's flexibility and portability, along with a gradient background to highlight the product's innovative design.

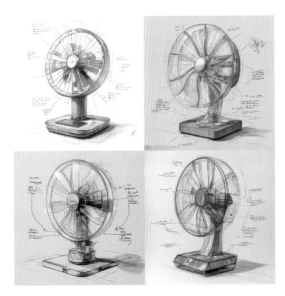

18. 節能環保風扇

A concept for an energy-efficient, eco-friendly fan made from recycled materials. The fan has a minimalist design with a transparent blade and a compact base. It operates silently and can be powered by solar energy. The concept is illustrated as a detailed hand-drawn sketch, with annotations highlighting its sustainable features.

19. 智能眼鏡

A concept for smart glasses that integrate augmented reality (AR) into everyday use. The glasses have a sleek, lightweight frame with a heads-up display that shows notifications, navigation, and real-time information. The design is modern and discreet, with a focus on blending technology seamlessly into daily life. The concept is illustrated in a digital render with realistic reflections and textures.

20. 變色水杯

A concept for a color-changing water bottle that reacts to temperature changes. The bottle is made from a special material that changes color depending on the temperature of the liquid inside. The design is sleek and ergonomic, with a matte finish. The concept is depicted in a simple digital illustration, focusing on the bottle's dynamic color transitions and user-friendly design.

01. 北歐風格客廳

A Scandinavian-style living room with light wood flooring, minimalist furniture, and large windows that allow natural light to flood the space. The design includes a cozy, neutral-colored sofa, a coffee table made from reclaimed wood, and simple, functional decor such as a floor lamp and a woven rug. The walls are painted in soft white, with a few black-and-white art pieces hanging for contrast.

02. 工業風格廚房

An industrial-style kitchen featuring exposed brick walls, stainless steel appliances, and open shelving made from reclaimed wood. The design includes a large center island with a concrete countertop, metal barstools, and pendant lights with exposed bulbs. The flooring is dark hardwood, and the overall color scheme is a mix of grays, blacks, and warm wood tones.

03. 美式鄉村風格臥室

A cozy American country-style bedroom with a rustic wooden bed frame, floral-patterned bedding, and a vintage quilt draped over the foot of the bed. The design includes a distressed wood dresser, an antique mirror, and bedside tables with ceramic lamps. The walls are painted in a soft, warm beige, and the floors are covered with a braided rug.

04. 極簡風格書房

A minimalist home office with a clean, uncluttered design. The room features a simple white desk with slim metal legs, a sleek ergonomic chair, and a single floating shelf on the wall holding a few carefully selected books and plants. The walls are painted in a soft, muted gray, and the flooring is light oak. A large window provides ample natural light, creating a serene and focused workspace.

05. 日式風格餐廳

A Japanese-inspired dining room featuring a low wooden table with floor cushions instead of chairs, creating a traditional tatami-style dining experience. The design includes sliding shoji screens, a bamboo pendant light, and a minimalistic centerpiece with a small bonsai tree. The walls are painted in a neutral beige, and the floor is covered with tatami mats. The overall ambiance is calm and serene, with a focus on simplicity and natural materials.

06. 現代藝術風格客廳

A modern art-inspired living room with bold colors and abstract patterns. The room features a large, bright blue sectional sofa, a geometric coffee table, and vibrant artwork on the walls. The design includes a mix of textures, such as a shaggy rug, metallic accents, and glass decor pieces. The walls are painted in a crisp white to allow the colors to stand out, and the overall vibe is energetic and eclectic.

07. 綠植室內花園

An indoor garden design with an emphasis on greenery and natural elements. The space features a variety of potted plants, vertical gardens, and a water feature with a small fountain. The design includes natural wood furniture, stone flooring, and large windows that let in plenty of sunlight. The background is kept minimal to focus on the lush, vibrant plants, creating a serene and refreshing environment.

08. 工業風格浴室

An industrial-style bathroom with exposed pipes, concrete walls, and a freestanding metal bathtub. The design includes a large round mirror with a metal frame, a wooden vanity with a stone countertop, and industrial-style lighting fixtures. The floor is made of dark, textured tiles, and the overall color palette is a mix of grays, blacks, and dark browns. The space has a raw, edgy feel with a touch of modern sophistication.

09. 美式現代風格廚房

An American modern-style kitchen with sleek, high-gloss cabinets, stainless steel appliances, and a large central island with a marble countertop. The design includes pendant lights with a modern twist, bar stools with upholstered seats, and a backsplash with subway tiles. The flooring is dark hardwood, and the color scheme is a mix of whites, grays, and metallics, giving the space a clean, contemporary look.

10. 地中海風格陽台

A Mediterranean-style balcony with terracotta tiles, wrought iron furniture, and vibrant blue and white accents. The design includes potted olive trees, a ceramic mosaic table, and colorful cushions. The walls are painted in warm earth tones, and the space is adorned with hanging lanterns and patterned floor tiles. The overall ambiance is relaxed and inviting, reminiscent of coastal Mediterranean homes.

11. 鄉村風格廚房

A rustic country-style kitchen featuring open shelving made from reclaimed wood, vintage-inspired appliances, and a large farmhouse sink. The design includes a wooden kitchen island with a butcher block countertop, copper pots hanging from a ceiling rack, and a brick backsplash. The flooring is wide plank wood, and the overall look is warm and inviting with a focus on natural materials.

12. 現代風格浴室

A modern-style bathroom with clean lines, a freestanding bathtub, and floor-to-ceiling glass shower enclosure. The design includes a floating vanity with a marble countertop, minimalist chrome fixtures, and a large mirror with built-in LED lighting. The walls are covered in large-format tiles in a neutral gray, and the floor is heated for added comfort. The space is sleek and luxurious, with a focus on functionality and aesthetics.

13. 洛可可風格客廳

A Rococo-style living room with ornate furniture, gilded mirrors, and pastel-colored walls. The design includes a vintage chandelier with crystal droplets, a tufted sofa with embroidered cushions, and intricate wall moldings. The flooring is dark hardwood, and the overall look is opulent and elegant, with an emphasis on elaborate details and luxurious materials.

14. 工業風格辦公室

An industrial-style office with exposed brick walls, metal beams, and large factory-style windows. The design includes a reclaimed wood desk, metal filing cabinets, and an open ceiling with visible ductwork. The lighting is provided by pendant lights with Edison bulbs, and the overall color scheme is a mix of grays, blacks, and natural wood tones. The space is functional and raw, with a modern edge.

15. 極簡風格臥室

A minimalist bedroom featuring a low platform bed with a simple wooden frame, white linen bedding, and a single potted plant as decoration. The design includes a small bedside table with a modern lamp, floor-to-ceiling windows with sheer curtains, and a neutral color palette of whites and grays. The flooring is light wood, and the space is uncluttered and serene, with a focus on simplicity and calm.

16. 波西米亞風格陽台

A Bohemian-style balcony with a mix of colorful textiles, patterned floor cushions, and hanging plants. The design includes a low wooden table with lanterns and candles, a hammock chair, and a macramé wall hanging. The floor is covered with a vibrant patterned rug, and the overall vibe is relaxed and eclectic, with an emphasis on comfort and personal expression.

17. 復古風格廚房

A retro-style kitchen featuring vintage appliances, checkered flooring, and pastel-colored cabinets. The design includes a classic diner-style booth, chrome accents, and a formica countertop. The backsplash is made of small square tiles in a vibrant color, and the lighting is provided by a retro pendant light. The overall look is nostalgic and cheerful, with a playful use of color and patterns.

18. 中式風格書房

A Chinese-inspired study with dark wood furniture, traditional calligraphy scrolls on the walls, and a large wooden desk with intricate carvings. The design includes a pair of tall bookshelves filled with classic literature, a low tea table with floor cushions, and a bonsai tree in the corner. The floor is covered with tatami mats, and the overall ambiance is calm and reflective, with a focus on cultural heritage.

19. 混搭風格客廳

An eclectic living room that combines elements from different design styles, including a mid-century modern sofa, a rustic coffee table, and an art deco chandelier. The design includes bold wallpaper, a mix of patterned throw pillows, and a gallery wall with a variety of art pieces. The flooring is dark wood, and the overall look is vibrant and unique, with an emphasis on personal style and creativity.

20. 斯堪地那維亞風格餐廳

A Scandinavian-inspired dining room with light wood furniture, simple lines, and a neutral color palette. The design includes a long wooden table with matching chairs, pendant lights with clean designs, and large windows that allow plenty of natural light. The flooring is light oak, and the walls are painted in soft white. The overall ambiance is airy and peaceful, with a focus on functionality and simplicity.

21. 洛可可風格臥室

A Rococo-style bedroom featuring a lavish canopy bed with intricate carvings, pastel-colored walls, and gilded accents. The design includes ornate mirrors, a chandelier with crystal droplets, and a plush velvet armchair. The flooring is dark wood, and the bedding is luxurious with layers of fine fabrics and embroidery. The overall ambiance is opulent and romantic, with a focus on elaborate details and rich textures.

22. 田園風格陽台

A cottage-style balcony with wooden furniture, floral-patterned cushions, and potted herbs. The design includes a small round table with a lace tablecloth, a wicker chair with a cozy throw, and hanging flower baskets. The floor is covered with natural stone tiles, and the walls are adorned with climbing plants. The overall look is quaint and charming, evoking the feel of a countryside retreat.

23. 現代風格客廳

A modern-style living room with a monochromatic color scheme, sleek furniture, and minimalist decor. The design includes a low-profile sectional sofa, a glass coffee table, and a large abstract painting as the focal point. The room is lit by recessed lighting and a floor lamp with a clean design. The flooring is polished concrete, and the overall look is sophisticated and contemporary, with an emphasis on open space and light.

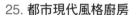

24. 新古典風格書房

A neoclassical study with dark wood paneling, a large mahogany desk, and classical sculptures. The design includes a leather wingback chair, floor-to-ceiling bookshelves filled with antique books, and a crystal chandelier. The walls are painted in a rich, deep green, and the floor is covered with a Persian rug. The overall ambiance is scholarly and refined, with a focus on timeless elegance and intellectual depth.

25. 都市現代風格廚房

An urban modern-style kitchen with sleek cabinetry, a large central island, and industrial lighting fixtures. The design includes stainless steel appliances, quartz countertops, and a subway tile backsplash. The flooring is polished concrete, and the overall look is clean and efficient, with an emphasis on functionality and contemporary aesthetics.

01. 火焰戰士

A fiery warrior with long, flame-colored hair and eyes that burn like embers. She wears a red and black battle suit embroidered with ancient flame symbols. Surrounding her body is a radiant aura of fire as she controls flames with her hands, standing determined on a battlefield. --ar 3:4

02. 未來戰士

A futuristic soldier with a single cybernetic eye that enhances his vision. He wears a sleek, armored suit with neon accents, and carries a high-tech plasma rifle. He stands in the middle of a devastated cityscape, scanning for threats with his enhanced vision. --ar 16:9

03. 未來探險家

A futuristic adventurer with a body partially made of advanced mechanical parts. His right arm and legs are metallic, glowing with a blue hue, and his left eye is a cybernetic lens. He wears a military-style suit with a large energy sword on his back, exploring ancient ruins in a desolate landscape. --ar 4:5

04. 天空守護者

A guardian of the skies with large, ethereal wings that resemble clouds. His hair is silver, and his eyes are deep blue, mirroring the sky. He wears a flowing white robe that glows faintly. He hovers above a city, watching over the inhabitants below, with the sky swirling with storm clouds behind him. --ar 3:4

05. 黑暗刺客

A shadowy assassin dressed in black from head to toe, with a mask covering his face. He wields dual blades that shimmer with a dark energy. His movements are swift and silent as he navigates through a dimly lit alley, his presence almost imperceptible. --ar 2:3

06. 龍魂法師

A mage who harnesses the power of ancient dragons. He has fiery red hair and golden eyes that glow with an inner fire. His robe is adorned with dragon scales, and he carries a staff topped with a dragon's head. He stands atop a mountain, summoning a colossal dragon from the clouds. --ar 4:5

07. 狐狸靈魂

A mystical being with the form of a silver fox, capable of transforming into a human. In her human form, she has long silver hair and piercing golden eyes. She wears traditional Japanese clothing adorned with fox motifs. She stands in a moonlit forest, surrounded by glowing spirits. --ar 3:4

08. 機械守護者

A massive, robotic guardian who protects an ancient temple. His body is made of interlocking metal plates, and his eyes glow with a blue light. He wields a large shield and a mace, standing imposingly at the entrance of the temple, ready to defend against intruders. --ar 4:5

09. 水之巫女

A serene water priestess with flowing blue hair and a dress made of water that ripples with her movements. She has the ability to control water in all its forms, from gentle streams to powerful waves. She stands at the edge of a tranquil lake, her hands raised as she summons water spirits. --ar 3:4

10. 影子舞者

A mysterious dancer who blends into the shadows. She has raven-black hair and wears a flowing black outfit that seems to merge with the darkness. Her movements are graceful and silent as she dances under a night sky filled with stars, her presence almost ethereal. --ar 2:3

11. 騎士公主

A noble princess clad in shining armor, holding a sword and shield emblazoned with her family's crest. Her golden hair flows beneath her helmet, and her eyes are filled with determination. She stands atop a castle wall, ready to defend her kingdom against invaders. --ar 3:4

12. 機械女僕

A futuristic robotic maid with a sleek, metallic body and glowing eyes. She carries a tray with various tools and gadgets, ready to assist her owner. Her movements are fluid and precise, embodying the perfect blend of technology and elegance. --ar 4:5

13. 龍騎士

A dragon rider clad in silver armor, soaring through the sky on the back of a fierce, winged dragon. His armor gleams in the sunlight, and he wields a lance that crackles with fire. The dragon's scales shimmer with iridescent colors as they dive towards their enemies. --ar 16:9

14. 雷霆戰士

A warrior wielding the power of lightning, with glowing blue eyes and armor that crackles with electric energy. He holds a massive hammer that sparks with electricity. The sky above him is stormy, with bolts of lightning striking the ground as he charges into battle. --ar 3:4

15. 神秘占卜師

A mystical fortune teller dressed in flowing robes adorned with ancient symbols. She has long, flowing black hair and deep purple eyes that seem to see into the future. Her hands hover over a crystal ball that glows with an inner light, revealing glimpses of possible futures. --ar 2:3

16. 雪之精靈

A graceful snow spirit with long, flowing white hair and pale blue skin that glimmers like ice. She wears a dress made of snowflakes that twinkle in the light. She dances through a winter forest, leaving a trail of frost in her wake. --ar 3:4

17. 邪惡巫師

A dark sorcerer cloaked in shadowy robes, with glowing red eyes and hands that crackle with dark magic. He stands in the center of a ritual circle, summoning dark forces to do his bidding. The air around him is thick with the scent of brimstone and the sound of chanting spirits. --ar 2:3

18. 未來忍者

A futuristic ninja equipped with advanced technology, wearing a suit that allows him to become invisible. His mask displays real-time data, and his weapons are sleek and high-tech, including a katana that can cut through steel. He moves silently through a neon-lit cityscape. --ar 16:9

19. 森林精靈

A woodland elf with emerald green eyes and hair that resembles leaves and vines. She is dressed in natural armor made from bark and leaves, blending seamlessly with the forest. She is seen standing in a grove, communicating with animals and guiding the growth of the forest. --ar 3:4

20. 星際探險家

An interstellar explorer in a sleek space suit, with a visor that reflects distant galaxies. She floats in the void of space, tethered to a spaceship, with a holographic map projected in front of her, charting new worlds to explore. --ar 16:9

01. 時間凍結的城市

A futuristic city where time has suddenly frozen. High-tech hovercars are suspended mid-air, people are caught mid-motion, and birds are frozen in flight. The entire scene is bathed in the golden glow of the sun, casting sharp shadows that are frozen in place. In the sky, a massive mechanical clock hovers, its hands stuck at zero, symbolizing the end of time or a new beginning. The atmosphere is eerie yet serene, as if the concept of time has been entirely removed from the world. --ar 16:9

02. 末日未來城市

A dystopian futuristic cityscape, where towering skyscrapers are crumbling, and the streets are filled with abandoned vehicles and overgrown vegetation. The sky is a sickly green, filled with smog and dark clouds. In the distance, a massive digital billboard still flickers with broken advertisements, creating an eerie, post-apocalyptic atmosphere. --ar 16:9

03. 恐怖的廢棄遊樂園

A terrifying scene of an abandoned amusement park at night, with broken rides and creepy, faded clown faces. The once joyful place is now overgrown with weeds and vines, with rusty chains swinging in the cold wind. The moonlight casts eerie shadows, and the distant sound of creaking metal adds to the chilling atmosphere. --ar 2:3

04. 神話中的奧林匹斯山

A grand depiction of Mount Olympus, home of the Greek gods, with towering marble temples surrounded by clouds. The gods are seen in majestic poses, with Zeus holding his lightning bolt, and Athena standing proudly with her shield and spear. The scene is filled with divine light, making the entire setting appear ethereal and powerful. --ar 3:2

05. 超現實夢境

A surreal dreamscape where gravity is defied, with floating islands, upside-down mountains, and rivers flowing upwards into the sky. The landscape is bathed in an otherworldly glow, with giant, glowing moons hanging low in the horizon. The scene is both mesmerizing and disorienting, blending reality with fantasy. --ar 4:5

06. 鬼怪的森林

A haunted forest scene where the trees are twisted and gnarled, with ghostly apparitions floating between them. The ground is covered in fog, and the moonlight barely penetrates the thick canopy. Shadows move independently of any visible source, creating a sense of dread and the feeling of being watched. --ar 3:4

07. 科幻機械都市

A sprawling futuristic city where towering skyscrapers are connected by massive mechanical bridges. The buildings are sleek and metallic, with neon lights illuminating the night. Giant robots patrol the streets, while flying vehicles zoom between the buildings. The atmosphere is bustling and energetic, representing the pinnacle of technological advancement. --ar 16:9

08. 未來感的虛擬世界

A futuristic virtual world where everything is made of neon grids and holographic structures. People are seen interacting with virtual interfaces, their avatars glowing in vibrant colors. The sky is a digital matrix, with data streams flowing like rivers. The atmosphere is high-tech and immersive, drawing the viewer into a world of endless possibilities. --ar 16:9

09. 神話中的天空之城

A mythical floating city in the clouds, supported by massive, ancient trees with roots that extend deep into the earth below. The city's architecture is a blend of classical temples and futuristic structures, with golden bridges connecting the different sections. The scene is bathed in golden sunlight, with birds flying between the floating islands. --ar 4:5

10. 超現實的倒影湖

A surreal scene of a lake with perfectly still water, where the reflections of the sky and surrounding mountains are clearer than reality itself. The surface of the water mirrors the sky in such detail that it becomes difficult to distinguish between the two. The horizon appears to vanish, creating an infinite loop of sky and water, blending reality with illusion. --ar 3:2

11. 未來的都市叢林

A futuristic cityscape where nature has reclaimed the urban environment. Skyscrapers are overgrown with vines and trees, and the streets are filled with lush vegetation. Animals roam freely in what was once a bustling metropolis. The scene is a harmonious blend of advanced technology and untamed nature, suggesting a new era of coexistence. --ar 16:9

12. 恐怖的陰暗地牢

A horror scene set in a dark, damp dungeon filled with rusty chains, skeletal remains, and eerie inscriptions on the walls. The air is thick with the stench of decay, and the only light comes from a flickering torch on the wall. Shadows seem to move on their own, and the distant sound of dripping water echoes through the corridors, amplifying the sense of dread. --ar 2:3

13. 神話中的龍之巢穴

A mythological scene of a dragon's lair hidden deep within a mountain. The lair is filled with gold, jewels, and ancient relics, all illuminated by the dragon's fiery breath. The dragon itself is massive, with scales that shimmer like molten metal. The atmosphere is tense and ominous, as if the dragon is guarding a great secret. --ar 3:2

14. 末日後的鬼怪廢墟

A post-apocalyptic wasteland where ruined buildings are haunted by ghostly figures. The landscape is barren, with twisted metal and broken concrete scattered everywhere. The sky is a dark red, filled with ash and embers, casting a hellish glow over the desolate scene. The spirits of the dead roam the ruins, creating an atmosphere of despair and dread. --ar 16:9

15. 太空中的巨大生物

A science fiction scene depicting a massive, bioluminescent creature drifting through the vastness of space. The creature's body is covered in glowing patterns, resembling a mix of deep-sea life and alien features. It moves gracefully among the stars, dwarfing nearby spacecraft. The scene is both awe-inspiring and unsettling, capturing the majesty and mystery of space. --ar 16:9

16. 超現實的無限樓梯

A surreal scene featuring an infinite staircase that spirals endlessly into the sky. The staircase is made of polished marble, with each step reflecting the strange, otherworldly light. People are seen ascending and descending the stairs, but they never reach an end or beginning, creating a sense of eternal loop and mystery. --ar 4:5

17. 鬼怪出沒的廢棄醫院

A haunted, abandoned hospital with broken windows, peeling walls, and medical equipment left to rust. The once bustling halls are now silent, with ghostly figures occasionally appearing in the corners of the room. The atmosphere is filled with a sense of unease and lingering spirits, making it a place where the boundary between the living and the dead is thin. --ar 3:4

18. 科幻的量子實驗室

A futuristic quantum laboratory where scientists are conducting experiments with particles that float and change shape in mid-air. The walls are made of transparent screens displaying complex data streams, and the air is filled with a faint glow from the quantum fields. The scene is cutting-edge and mysterious, representing the frontier of scientific discovery. --ar 16:9

19. 超現實的冰山宮殿

A surreal ice palace set within a colossal iceberg floating in the middle of a vast, frozen ocean. The palace is made entirely of crystalline ice, with grand halls and intricate sculptures. Light refracts through the ice, casting rainbows across the walls. The atmosphere is cold and ethereal, as if the palace exists in a different dimension. --ar 3:2

20. 未來感的沉浸式遊戲世界

A futuristic, immersive gaming world where players are fully integrated into the game environment. The landscape is a blend of hyper-realistic graphics and abstract digital elements, with players navigating through ever-changing terrains and battling AI-controlled creatures. The scene is vibrant and interactive, representing the future of virtual reality entertainment. --ar 16:9

21. 神話中的地獄之門

A mythological scene of the Gates of Hell, where massive, iron doors are embedded in a mountain of fire and brimstone. The ground is cracked and lava flows freely, with tortured souls trapped in the rock walls. The sky is a swirling vortex of dark clouds and lightning, and the atmosphere is one of eternal damnation and suffering. --ar 4:5

22. 無重力的夢境森林

A surreal, zero-gravity forest where trees, plants, and animals float freely in the air. The roots of towering trees dangle in mid-air, intertwined with floating rocks, while vibrant flowers drift like glowing creatures. Animals move gracefully through the air as if swimming, and a river flows through the sky, its water shimmering with lights like stars. The sky is filled with soft pink and blue mists, adding to the mysterious and dreamlike atmosphere. --ar 4:5

01. 蒙古族 (Mongolian)

A Mongolian man in traditional Deel clothing, wearing a thick fur hat and a belt with traditional patterns. He stands in the vast grasslands with yurts and grazing horses in the background, holding an eagle. Photographed in the style of Steve McCurry, with vibrant colors and natural light highlighting the rugged texture of his attire and the expansive landscape. --ar 3:2

02. 苗族 (Miao)

A Miao woman in a silver headdress and embroidered clothing with intricate patterns and symbols. She is photographed during a festival in a mountainous landscape, surrounded by other villagers. The image is captured with soft, diffused light, creating a dreamy atmosphere. The composition is balanced, with the woman as the central focus. --ar 4:5

03. 維吾爾族 (Uyghur)

A Uyghur man in a traditional doppa hat and long robe with detailed embroidery. He stands in the bustling bazaar of Kashgar, surrounded by market stalls selling spices, fruits, and textiles. Shot in black and white with high contrast, reminiscent of Sebastião Salgado's documentary style, emphasizing the textures and patterns of the scene. --ar 16:9

04. 藏族 (Tibetan)

A Tibetan woman in traditional attire, including a chuba (long-sleeved robe) and a wide-brimmed hat adorned with turquoise and coral jewelry. She stands in front of a backdrop of snow-capped mountains and prayer flags fluttering in the wind. The photograph is taken in golden hour light, enhancing the warm tones of her clothing and the landscape. --ar 3:2

05. 侗族 (Dong)

A Dong man dressed in traditional indigo-dyed cotton clothing, featuring intricate embroidery and silver ornaments. He is seated in front of an ancient wooden stilt house in his village, with a peaceful river flowing in the background. The image is captured in a documentary style with natural light, emphasizing the cultural details and the serene environment. --ar 3:2

06. 傣族 (Dai)

A Dai woman wearing a traditional peacock-patterned dress, standing near a lush green bamboo grove. Her attire is colorful and features intricate patterns that represent nature and fertility. The photo is captured in soft morning light, with a shallow depth of field to focus on her serene expression. --ar 4:5

07. 哈尼族 (Hani)

A Hani woman in traditional black and red garments, including a headdress adorned with silver coins. She stands in a terraced rice field, the curves of the terraces leading the eye towards the horizon. The photo is taken at dusk with soft, diffused light, creating a calm and balanced composition. --ar 16:9

08. 朝鮮族 (Korean Ethnic Minority in China)

A Korean ethnic minority woman in a traditional hanbok with vibrant colors, standing in front of a historic pagoda. The hanbok features delicate embroidery and a wide, flowing skirt. The photograph is shot with a wide-angle lens, capturing both the details of her attire and the grandeur of the architecture. --ar 4:5

09. 羌族 (Qiang)

A Qiang man wearing traditional woolen garments and a felt hat, standing on a stone-paved village path with mountains in the background. His attire is simple yet practical, with earth tones that blend into the surrounding landscape. The image is captured in a natural light setting, using a high-contrast black and white filter to emphasize texture and contrast. --ar 3:2

10. 祖魯族 (Zulu)

A Zulu woman in traditional attire, including a brightly colored beaded necklace and headband. She is photographed in her rural village, with thatched huts and lush vegetation surrounding her. The photo is taken with natural light, capturing the vivid colors and intricate patterns of her clothing. --ar 4:5

11. 彝族 (Yi)

A Yi woman in a traditional outfit with bold geometric patterns and vibrant colors, standing in front of a thatched-roof house. Her attire includes a large, colorful headscarf and layered skirts. The photograph is taken at midday, with strong sunlight casting sharp shadows that highlight the patterns and textures of her clothing. --ar 16:9

12. 馬賽族 (Maasai)

A Maasai warrior dressed in traditional red shuka cloth, adorned with beaded jewelry and carrying a spear. He stands tall in the open savannah, with acacia trees and a distant mountain range in the background. Photographed during the golden hour, the warm light emphasizes the vibrant colors and textures of his attire. --ar 3:2

13. 亞馬遜雨林部落 (Amazon Rainforest Tribe)

A member of an Amazonian tribe wearing body paint and a feathered headdress. He stands in the dense rainforest, surrounded by towering trees and vibrant green foliage. The image is captured in a documentary style, using natural light to emphasize the raw beauty and connection to nature. --ar 16:9

14. 馬庫西族 (Makushi)

A Makushi woman in traditional attire, including a beaded necklace and a colorful woven skirt, standing in a lush rainforest clearing. She is surrounded by native plants and carries a woven basket filled with fruits. The photograph is taken in soft, natural light, emphasizing the vibrant colors and intricate details of her clothing. --ar 4:5

15. 瑪雅族 (Maya)

A modern-day Maya woman in traditional huipil, a colorful handwoven blouse with intricate designs, standing in front of an ancient Mayan temple. She wears a vibrant headscarf and carries a handmade basket. The photo is taken at dawn, with soft light illuminating the scene, emphasizing the connection between the past and present. --ar 4:5

16. 桑比亞族 (Samburu)

A Samburu warrior dressed in traditional shukas, adorned with beaded jewelry and carrying a spear. He stands in front of his village with thatched huts and a large acacia tree in the background. The image is shot in the early morning, capturing the cool tones and long shadows of the savannah. --ar 3:2

17. 辛巴族 (Himba)

A Himba woman with traditional ochre-covered skin and elaborate braided hair, standing in front of her family's hut. She wears a goatskin skirt and intricate beaded jewelry. The photo is taken at midday, with harsh sunlight casting dramatic shadows, highlighting the rich textures of her skin and attire. --ar 3:2

18. 恩貝拉-沃南族（Emberá-Wounaan）

A Emberá-Wounaan man with intricate body paint and a feathered headdress, standing beside a river in the dense jungle. He is holding a handmade canoe paddle, with the river and lush greenery reflected in the water. The photograph is taken at midday, with high contrast to emphasize the patterns of his body paint and the surrounding environment. --ar 16:9

19. 卡耶波族 (Kayapo)

A Kayapo warrior wearing a large feathered headdress and body paint in geometric patterns. He stands proudly in front of a large communal hut, surrounded by other villagers. The photo is taken with a wide-angle lens to capture the scale of the scene and the intricate details of the warrior's attire. --ar 16:9

20. 富拉尼族 (Fulani)

A Fulani woman with a traditional hairstyle adorned with silver ornaments, wearing a vibrant indigo-dyed cloth wrapped around her body. She is seated in front of a mud-brick house, with a desert landscape in the background. The photograph is captured in warm evening light, with soft shadows highlighting the textures and colors of her attire. --ar 4:5

21. 貝都因族 (Bedouin)

A Bedouin man dressed in traditional flowing robes and a keffiyeh, standing beside his camel in the vast desert. The sun is setting, casting long shadows and a warm, golden hue over the sand dunes. The photo is captured in a cinematic style, emphasizing the expansive and harsh desert environment. --ar 16:9

22. 哈扎比族 (Hadza)

A Hadza hunter holding a bow and arrow, dressed in simple animal hides, standing in the Tanzanian savannah. The background features tall grasses and distant mountains under a clear blue sky. The photograph is taken with natural light, capturing the raw and unfiltered connection to nature. --ar 16:9

23. 韃靼族 (Tatar)

A Tatar woman in traditional dress, including an embroidered vest and a colorful headscarf, standing near a wooden mosque. She holds a basket of traditional bread. The photograph is taken in late afternoon light, with warm tones accentuating the textures of her clothing and the wood of the mosque. --ar 3:2

24. 奧莫河谷族 (Omo Valley Tribe)

A member of the Omo Valley tribe, adorned with body paint and elaborate beadwork, standing against the backdrop of the Ethiopian savannah. The person's hair is styled with clay and adorned with colorful beads. The photograph is taken with soft, diffused morning light, capturing the intricate details and vibrant colors. --ar 4:5

25. 伊博族 (Igbo)

An Igbo woman in a traditional wrapper and gele headtie, standing in front of a thatched-roof house. She carries a clay pot on her head with graceful balance. The photo is taken in the early morning, with soft light highlighting the intricate patterns of her attire. --ar 4:5

26. 克丘亞族 (Quechua)

A Quechua woman in traditional Andean clothing, including a brightly colored skirt and shawl, standing on a terraced field in the mountains of Peru. She is holding a spindle and wool, with the Andes mountains in the background. The photo is taken at midday, with the bright sunlight bringing out the vivid colors of her clothing. --ar 3:2

27. 因努伊特族 (Inuit)

An Inuit elder wearing a traditional fur parka with a large hood, standing on the frozen tundra with the Arctic landscape in the background. The elder holds a harpoon, with distant icebergs and a low sun casting a cold, bluish light over the scene. The photograph is taken in a high-contrast black-and-white style. --ar 3:2

28. 巴塔克族 (Batak)

A Batak elder in traditional clothing, including a woven ulos cloth draped over his shoulders, standing in front of a traditional Batak house with steep roofs. He is holding a wooden staff. The photograph is taken in the late afternoon, with warm light emphasizing the deep textures of his face and clothing. --ar 4:5

29. 馬普切族 (Mapuche)

A Mapuche woman in traditional clothing, including a poncho and silver jewelry, standing in front of a thatched-roof house in the Chilean countryside. She holds a wooden staff adorned with carvings. The photo is taken at dawn, with soft, warm light illuminating the scene, capturing the deep connection to her heritage. --ar 4:5

31. 哈扎拉族 (Hazara)

A Hazara man in traditional Afghan clothing, including a pakol hat and a woolen shawl, standing in a mountainous region of central Afghanistan. He is holding a prayer bead necklace, with snow-capped peaks in the distance. The photograph is taken in soft, diffused light, emphasizing the serene and spiritual atmosphere of the setting. --ar 4:5

30. 蘇魯族 (Sulu)

A Sulu warrior in traditional attire, including a brightly colored malong (wraparound skirt) and a kris sword at his side. He stands in a coastal village in the Philippines, with traditional stilt houses and palm trees in the background. The photo is taken with vibrant colors and natural light, capturing the dynamic culture of the Sulu people. --ar 3:2

32. 卡拉莫瓊族 (Karamojong)

A Karamojong elder wearing traditional beadwork and a patterned blanket draped over his shoulders, standing in front of a mud-and-thatch hut in northeastern Uganda. He holds a ceremonial staff adorned with cowrie shells. The photograph is taken at sunset, with the warm light casting long shadows, creating a sense of depth and history. --ar 3:2

33. 伊富高族 (Ifugao)

An Ifugao warrior wearing traditional attire, including a woven bahag (loincloth) and a headdress made of feathers and bones. He stands on a terraced rice field in the Philippines, with mist rising from the valley below. The photograph is taken with a low-angle perspective, emphasizing the warrior's strength and the majesty of the landscape. --ar 16:9

34. 恩德貝萊族 (Ndebele)

A Ndebele woman in traditional attire, including a beaded apron and colorful neck rings, standing in front of a brightly painted house. The patterns on her clothing mirror the geometric designs of the house behind her. The photograph is taken in bright daylight, with strong, vivid colors highlighting the cultural significance of the patterns. --ar 4:5

35. 阿散蒂族 (Ashanti)

An Ashanti king wearing a kente cloth with bold, colorful patterns, seated on a traditional wooden stool in front of a palace in Ghana. He is adorned with gold jewelry and a crown, symbolizing his royal status. The photograph is taken in natural light, with a focus on the richness of the textiles and the grandeur of the setting. --ar 3:2

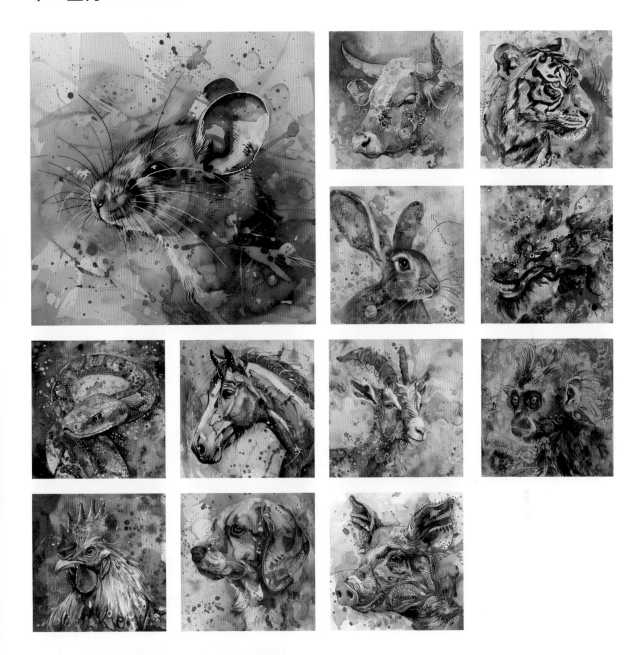

A vibrant and abstract watercolor painting of a {{mouse, bull, tiger, rabbit, chinese dragon, snake, horse, goat, monkey, rooster, dog, piggy}}, featuring bold and colorful splashes in the style of modern art. It is depicted in a front profile, with intricate patterns and vivid colors blending together to create a dynamic and expressive image. The background is a mix of vivid colors hues, adding depth and contrast to its multicolored fur. The overall style is whimsical and energetic, capturing the essence of it with a playful and artistic flair.

一幅充滿活力和抽象的 {{鼠、牛、虎、兔、龍、蛇、馬、羊、猴、雞、狗、豬}} 水彩畫，以現代藝術風格大膽多彩的潑濺為特色。它以正面輪廓描繪，複雜的圖案和鮮豔的色彩融合在一起，創造出充滿活力和表現力的圖像。背景是鮮豔色彩的混合色調，增加了其多彩皮毛的深度和對比。整體風格異想天開、充滿活力，以俏皮和藝術的風格抓住了它的精髓。-

《山海經》是先秦時期的古籍，其中記載許多民間傳說的妖怪，學者也用以考證奇物異俗，山川形勢，是一本具有歷史價值的著作。過去也有不少創作者根據書中記載編繪山海經圖錄，這裡我也嘗試用 AI 還原古人的智慧。

01. 夔牛 (Kui Niu)

A massive, one-legged beast resembling a bull, with a single spiraling horn atop its head. The creature's rough, dark skin is covered in thick hair, and it has a glowing red eye. The scene is set in a stormy landscape, depicted in a traditional Chinese ink wash painting style.

02. 九尾狐 (Nine-Tailed Fox)

A mystical fox with silver fur and nine flowing tails. Its deep blue eyes exude mystery. The fox moves gracefully in a moonlit forest, illustrated in a soft, ethereal digital painting style.

03. 青龍 (Azure Dragon)

A majestic dragon with a serpentine body covered in shimmering blue-green scales. It has golden antlers and sharp claws. The dragon soars above a lush landscape with cherry blossoms, depicted in a traditional Chinese watercolor style.

04. 燭陰 (Zhuyin)

A colossal dragon-like creature with a serpentine body covered in smooth red scales. Its glowing eyes can control light and darkness. The scene shows it hovering above jagged mountains in a dramatic fantasy style.

05. 白澤 (Bai Ze)

A white beast with a lion-like appearance, standing proudly on a mountain peak. Its eyes are deep and wise. The scene is in a snow-covered landscape, depicted in a traditional ink wash painting style.

06. 騶虞 (Zou Yu)

A five-colored tiger-like beast prowling through an ancient forest. Its fur is a mix of vibrant colors—red, blue, green, yellow, and white. The scene is illustrated in a high-contrast digital painting style.

07. 雷獸 (Lei Shou)

A thunderous beast with a tiger-like body and bird-like wings, surrounded by dark storm clouds. Its fur is striped, and it has sharp claws and fangs. The scene is captured in a powerful charcoal and pastel illustration.

08. 鯤鵬 (Kun Peng)

A giant mythical fish that transforms into a bird, with a massive body covered in dark, shimmering scales. When in its bird form, its wings stretch wide, with feathers that transition from deep blue at the base to golden tips. The scene depicts its transformation mid-air, above a turbulent ocean, in a surrealist painting style.

09. 狍鴞 (Pao Xiao)

A creature with the body of a deer and the face of an owl, with large, glowing eyes and a body covered in brown, spotted fur. Its antlers branch out like twisted trees. The scene is set in a misty, enchanted forest, illustrated in a whimsical, fairy-tale style.

10. 三足烏 (Three-Legged Crow)

A legendary crow with three legs, its feathers jet black with a metallic sheen. The crow's eyes are a bright, burning orange. The scene depicts it flying across the sky at sunset, with the sun casting a golden hue over the clouds, in a minimalist Japanese ink wash style.

11. 玄武 (Xuanwu)

A giant tortoise with a serpent coiled around its body, both creatures depicted in intricate jade carving. The tortoise's shell is dark green with ancient patterns, and the serpent's scales are finely detailed. The scene is set in a tranquil, flowing stream, illustrated in a traditional Chinese jade sculpture style.

12. 睚眥 (Yazi)

A fierce dragon-like creature with a tiger-like body and a lion's mane. Its scales are dark, with a hint of crimson, and its teeth are sharp and numerous. The scene shows it in a battle stance, roaring on a rocky battlefield, depicted in a vibrant digital painting style.

13. 朱厭 (Zhuyan)

A beast with a human-like face and the body of a lion, with reddish fur and a pair of large, forward-facing horns. Its eyes are dark and menacing. The scene is set in a dark, dense forest, captured in a traditional Chinese brush painting style with strong ink strokes.

14. 燭龍 (Zhulong)

A massive red dragon with a long, serpentine body and smooth scales that shine like molten lava. Its eyes are bright, illuminating the sky. The scene shows the dragon coiled around a mountain peak, with the sky split between day and night, depicted in an epic, high-fantasy illustration style.

15. 九頭鳥 (Jiu-Tou-Niao)

A mythical bird with nine heads, each head resembling different birds such as an eagle, owl, and crane. The feathers are multicolored, with each head displaying a unique pattern. The scene shows the bird soaring above a mystical landscape, illustrated in a modern, surrealist painting style.

16. 檮杌 (Taowu)

A ferocious creature with the body of a tiger and the face of a snarling demon. Its fur is thick and mottled, with dark stripes running across its muscular frame. The creature's face is grotesque, with large fangs and red eyes. The scene depicts it mid-attack in a dense jungle, illustrated in an ancient Chinese woodblock print style.

17. 巴蛇 (Ba She)

A giant serpent capable of swallowing entire elephants, with a long, sinuous body covered in dark, scaly skin. Its eyes are narrow and venomous, and its fangs are long and sharp. The scene shows the serpent coiled around a massive tree in a dense jungle, depicted in a cinematic, high-contrast digital painting style.

18. 獙狼 (Ao Hen)

A fierce dog-like creature with fiery red fur and sharp teeth. Its eyes burn with an intense fire, and its claws are long and curved. The creature is depicted standing on a rocky outcrop, with flames rising around it. The scene is illustrated in a traditional Chinese fire painting style, with vibrant reds and oranges dominating the palette.

19. 犰狳 (Qiu Yu)

A creature resembling a pangolin with a hard, armored shell covered in metallic scales. Its claws are sharp, designed for digging through tough earth, and its long tongue flicks out to capture prey. The scene shows it burrowing through rocky terrain, depicted in a detailed, realistic pencil sketch style.

20. 驪山 (Li Shan)

A black dragon with a long, serpentine body and smooth, glossy scales that reflect the surrounding light. Its eyes are a piercing yellow, and it has long, curved horns. The scene shows the dragon coiled around a mountain peak, with mist rising around it, depicted in a traditional Chinese ink and wash painting style.

01. 女媧補天

A sculpture concept of Nüwa mending the sky with five-colored stones, crafted from porcelain and hand-painted with delicate details. Nüwa is shown in a serene pose, carefully placing a stone in the sky, while the heavens above her shimmer with light. The base of the sculpture features swirling clouds and water motifs.

02. 哪吒鬧海

A vibrant depiction of Nezha fighting the Dragon King, illustrated in a traditional Chinese woodblock print style. Nezha is shown riding on a flaming wheel, with his spear aimed at the dragon. The waves of the ocean and the dragon's scales are intricately detailed, with bright, bold colors that capture the intensity of the battle.

03. 孫悟空大鬧天宮

A concept for an animated scene of Sun Wukong rebelling in Heaven, rendered in a modern 3D animation style. Sun Wukong is shown with his golden staff, battling against celestial soldiers amidst the clouds. The scene is dynamic and full of action, with vibrant colors and fluid movements that reflect the chaos of the battle.

04. 白蛇傳

A romantic and mysterious depiction of the Legend of the White Snake, illustrated in a classical Chinese watercolor style. The White Snake is shown in human form, with delicate features and flowing robes, standing beside Xu Xian on a bridge. The scene is misty, with soft, muted colors, and the background features a serene West Lake with distant pagodas.

05. 精衛填海

A stylized charcoal drawing of Jingwei filling the sea, illustrated in a minimalistic, symbolic art style. Jingwei is depicted as a small bird determinedly carrying a stone towards the vast ocean. The waves are drawn with bold, sweeping lines, and the overall composition emphasizes perseverance and determination.

06. 齊天大聖

A vibrant, action-packed depiction of Sun Wukong (Monkey King) in his role as the Great Sage Equal to Heaven, rendered in a dynamic comic book style. Sun Wukong is shown wielding his golden staff, with his iconic cloud somersaults in the background. The scene is filled with energy, using bold lines and vivid colors to capture his rebellious spirit.

07. 梁祝化蝶

A romantic and tragic illustration of Liang Shanbo and Zhu Yingtai transforming into butterflies, illustrated in a delicate, pastel-colored art style. The scene captures the moment of transformation, with the lovers' figures gracefully merging into two butterflies amidst a field of blooming flowers. The colors are soft and muted, evoking a sense of both beauty and sorrow.

行筆至此，其實心中還是感觸良多的，我曾經在一些場合表示，未來 AI 模型的訓練也彰顯了文化的強弱，目前看來是朝這樣發展——AI 的話語權。例如同樣都是神話傳說，我們來比較中西方的差異。

想法 ⇒ 輸入：特洛伊戰爭。

Trojan War --ar 16:9 --style raw

想法 ⇒ 輸入：封神演義

Fengshen Yanyi --ar 16:9

是有點不甘心的，所以有了接下來的一些中國神話故事創作，當然更期待有一天，我可以直接輸入「桃園三結義」或是「赤壁之戰」就可以生成適當的影像。

繪本創作 LOOK 🖥 495

類似的主題，可以用不同的媒材及風格來表現，也可以用同樣的風格，來創作一系列的作品。以下我們以橫尾忠則詭譎多變、具強烈色彩的迷幻風格，搭配適當的主題描述，創作一系列的中國神話故事。在之前我們不斷提及目前歐美的影像模型對於東方題材的訓練數是不足的，也因此我們無法很單純地在提示詞中加入某個神話人物的姓名就生成相對應的主題背景，而是必須要提供更多的敘事內容，也不妨將此視為撰寫提示詞的訓練。而透過這樣單一風格，不同主題內容，便可應用在諸如繪本創作上。

01. 嫦娥奔月

An illustration in the style of Tadanori Yokoo depicting Chang'e, the Chinese moon goddess, ascending to the moon. The scene features Chang'e with a mystical and determined expression, surrounded by bold, psychedelic patterns and vibrant colors. The art style includes Yokoo's signature use of surreal and eclectic elements, with intricate designs, striking contrasts, and a sense of dynamic movement. The background showcases a fantastical moonlit sky with stars and a glowing moon, interwoven with abstract and symbolic motifs. --ar 3:2 --style raw "

02. 神農嘗百草

An illustration in the style of Tadanori Yokoo depicting Shennong tasting herbs. The scene features Shennong with a contemplative and determined expression, surrounded by a variety of vibrant and mystical plants. The art style includes Yokoo's signature use of bold, psychedelic patterns and vibrant colors, with intricate designs and surreal elements. The background showcases a fantastical landscape with swirling colors, abstract forms, and symbolic motifs, capturing the mythic nature of Shennong's quest to discover medicinal herbs. --ar 3:2 --style raw

03. 盤古開天

An illustration in the style of Tadanori Yokoo depicting Pangu creating the heavens and the earth. The scene features Pangu with a powerful and determined expression, wielding a massive axe as he separates the sky from the earth. The art style includes Yokoo's signature use of bold, psychedelic patterns and vibrant colors, with intricate designs and surreal elements. The background showcases a chaotic and dynamic landscape with swirling clouds, cosmic motifs, and abstract forms, capturing the mythic moment of creation. --ar 3:2 --style raw

盤古開天
Pangu Genesis, by Louis BZ Huang Creations

千里眼順風耳
Clairvoyant and Clairaudient, by Louis BZ Huang Creations

04. 千里眼順風耳

An illustration in the style of Tadanori Yokoo depicting Clairvoyant (Qianliyan) and Clairaudient (Shunfeng'er). The scene features Qianliyan with a piercing gaze, holding a spyglass, and Shunfeng'er with large ears, intently listening, each displaying their extraordinary abilities. The art style includes Yokoo's signature use of bold, psychedelic patterns and vibrant colors, with intricate designs and surreal elements. The background showcases a fantastical landscape with swirling colors, abstract forms, and symbolic motifs, capturing the mystical and supernatural essence of their powers. --ar 3:2 --style raw

05. 大禹治水

An illustration in the style of Tadanori Yokoo depicting Yu the Great controlling the floods. The scene features Yu with a focused and determined expression, using his tools to redirect the water flow and tame the floods. The art style includes Yokoo's signature use of bold, psychedelic patterns and vibrant colors, with intricate designs and surreal elements. The background showcases a dramatic landscape with swirling water, abstract forms, and symbolic motifs, capturing the mythic and heroic nature of Yu's efforts to control the floods. --style raw --ar 3:2 --v 6.0

大禹治水
Dayu controlled floods, by Louis BZ Huang Creations

女媧造人
Nuwa Creating Humans, by Louis BZ Huang Creations

06. 女媧造人

An illustration in the style of Tadanori Yokoo depicting Nüwa creating humans. The scene features Nüwa with a serene and focused expression, holding clay figures that are transforming into humans. The art style includes Yokoo's signature use of bold, psychedelic patterns and vibrant colors, with intricate designs and surreal elements. The background showcases a mystical landscape with swirling colors, abstract forms, and symbolic motifs, capturing the divine and creative essence of the scene. --ar 3:2 --style raw

07. 牛郎織女

An illustration in the style of Tadanori Yokoo depicting the myth of the Cowherd and the Weaver Girl (Niulang and Zhinu). The scene features Niulang and Zhinu with longing and affectionate expressions, reaching out towards each other across the Milky Way. The art style includes Yokoo's signature use of bold, psychedelic patterns and vibrant colors, with intricate designs and surreal elements. The background showcases a celestial landscape with swirling stars, abstract forms, and symbolic motifs, capturing the romantic and mythical essence of their love story. --ar 3:2 --style raw

08. 后羿射日

An illustration in the style of Tadanori Yokoo depicting Hou Yi shooting down the suns. The scene features Hou Yi with a powerful and determined expression, drawing his bow with a glowing arrow aimed at one of the multiple suns in the sky. The art style includes Yokoo's signature use of bold, psychedelic patterns and vibrant colors, with intricate designs and surreal elements. The background showcases a fiery sky with several suns, surrounded by abstract forms and symbolic motifs, capturing the mythic and heroic nature of the scene. --ar 3:2 --style raw

09. 夸父逐日

An illustration in the style of Tadanori Yokoo depicting Kuafu chasing the sun. The scene features Kuafu with a determined and intense expression, running towards a large, radiant sun in the sky. The art style includes Yokoo's signature use of bold, psychedelic patterns and vibrant colors, with intricate designs and surreal elements. The background showcases a dynamic landscape with swirling abstract forms, symbolic motifs, and a fiery sky, capturing the mythic and relentless pursuit of the sun. --style raw --v 6.0 --ar 3:2

▶ 來幫孩子做一本故事書吧！

敘事創作

❶ 先設定一個主題與人物：描述主角的外貌，包括年紀、種族、服飾等等，在這個範例，我們設定的是「**一個五歲的美國女孩，金髮，穿著花色洋裝。**」

❷ 再設定插畫風格：為「**新海誠、日本動漫風格**」。

❸ 根據設定的故事情節**給予場景、動作、表情的設定**，生成每頁獨立的圖像後，在透過一些編輯工具加入文字即可。

使用角色參照參數，接續後續創作

❶ 主角是一隻小獅子，我們先生成一隻小獅子的圖案，採用簡單的線條、平面插畫風格。

Cute lion, simple lines, flat illustration style, white background, yellow and orange tones, children's book illustration style, simple details, hand-drawn texture, cute expression, watercolor, cartoon character design, solid color blocks, white space composition, simple background, full-body portrait, high definition. --ar 64:51 --style raw

❷ 然後我們可以將這上面這張圖上傳至一個網路空間，記下圖片連結，便可使用 "—cref (image url)"角色參照參數，接續後續的創作。跟前一個例子相同，只需要增加背景、動作或是其他故事元素即可。

A cozy lion's den nestled within a lush jungle. The den is warm and inviting, with soft earthy tones and sunlight streaming in through the foliage. The little lion, with golden fur and a determined expression, stands at the entrance, ready to embark on its food adventure. flat illustration style, simple lines, bright colors, simple details, --ar 5:4 --style raw --niji 6 --s 750 --cref https://s.mj.run/_aouRL8PaHk

到了這裡，我相信大家應該都會覺得用 AI 進行創作是件容易的事，從另外一個角度思考也就是門檻降低，差異性也降低，於是乎最重要的就變成是創意而非技巧，而現在則具備了將我們的創意實現落地的能力，在可期待的未來，這部分的能力會更快、更好。

Runway、Sora 讓圖片動起來！

除了靜態影像外，本書所提到的提示詞敘述邏輯，基本上都可用在任何 AI 生成應用上，當然也包括動畫及影音檔生成，例如最近很火熱的 Sora，或是之前很多人使用的 Runway 等等。除了可直接使用提示詞生成外，這些影片生成軟體，都有提供以靜態影像生成動畫影片的功能。

生成範例│眨眼微笑打招呼

❶ 先使用 Midjourney 生成一張圖。

A wide-angle, top-down view featuring a happy 30-year-old bald Asian man wearing glasses, standing proudly in front of a customized Porsche 911 GT3 converted into a van. The entire vehicle is clearly visible from this bird's-eye perspective, showcasing the unique design and modifications. The background features a car modification workshop, with visible tools, car parts, and equipment scattered around, adding to the industrial atmosphere. The lighting highlights both the man's joyful expression and the sleek lines of the car from this dynamic aerial angle. --ar 3:2

❷ 運用 P.000 提到的小工具執行換臉動作。

❸ 在 runway 平台（https://app.runwayml.com）加上動作的提示：man turn around and smile to the camera.

生成範例│走進亞特蘭提斯海底王國

A cinematic underwater scene depicting the mythical lost city of Atlantis. Show a grand, submerged metropolis with towering ancient temples, mysterious glowing runes, and colossal statues of forgotten deities. Include diverse marine life like shimmering schools of fish, majestic whales, and bioluminescent creatures illuminating the dark ocean depths. The atmosphere should be mystical and awe-inspiring, with cinematic lighting that evokes a sense of adventure and epic storytelling. Use dynamic camera angles, soft blue and turquoise tones, and an ethereal glow that creates a sense of wonder and mystery. --ar 16:9 --style raw --v 6.1

❷ 在 sora 平台（https://openai.com/sora）的生成頁面上輸入相同的提示詞。

生成範例│阿里山日出與群鳥飛飛

❶ 取 P.436 的「16. 阿里山日出圖」。

Ukiyo-e showcasing the breathtaking sunrise over Alishan. The composition features the iconic sea of clouds with the sun rising above the peaks, casting a warm, golden glow over the forested landscape. The distant mountains are softly outlined, creating a peaceful atmosphere. In the style of Katsushika Hokusai. --ar 3:2

❷ 在 Sora 的生成介面上,加上一段提示詞:一群鳥在飛。

A group of birds flying

再將 Sora 生成的影片檔,透過編輯軟體合成音效,下面這個範例使用的是 canva.com。

透過這幾個範例,希望讀者更能理解,生成式 AI 的邏輯,並不在於動態或靜態,核心的創作邏輯還是來自於對於生成主題的精確描述。

觀音大士心經

AI 工具城市介紹

● AI 寫作生成器

工具程式	連結處	支援平台	補充說明
Jasper	https://www.jasper.ai	網頁版	提供多樣的文案生成範本與可客製化的品牌風格，幫助企業維持一致的品牌形象。
Anyword	https://anyword.com	網頁版	擅長一鍵生成廣告文案、社群貼文、標題等行銷內容，並提供 A/B 測試與成效分析。
Shortwave	https://www.shortwave.com	網頁版（整合 Gmail）	專為 Gmail 用戶打造，透過 AI 幫助撰寫郵件、整理信箱，讓回信更快速。

● AI 筆記

工具程式	連結處	支援平台	補充說明
Notion AI	https://www.notion.so/product/ai	網頁版、桌面版、行動 App	整合於 Notion 筆記系統，可自動摘要、生成內容、翻譯等，讓資訊管理更輕鬆。
Mem	https://get.mem.ai/	網頁版、行動 App (iOS)	聚焦在個人知識管理，能用 AI 快速記錄、搜尋與整理，保持筆記有條不紊功能。

● AI 影片工具

工具程式	連結處	支援平台	補充說明
Runway	https://runwayml.com	網頁版	提供影片後製、特效生成與剪輯的一站式解決方案，支援多種 AI 視覺處理技術。
Wondershare Filmora	https://filmora.wondershare.com	桌面版（Windows/Mac）	結合 AI 與易用介面，可自動優化影片剪輯節奏、轉場特效等，迅速完成影片編輯。
Sora	https://openai.com/sora	網頁版	最新的 AI 影片生成工具，可根據文字或素材自動產生影片內容，應用範圍多元。

● AI 影像工具

工具程式	連結處	支援平台	補充說明
Midjourney	https://midjourney.com	透過 Discord、網頁版	在 Discord 下指令即可快速生成高質感圖像，適合各類設計、美術創作。
DALL·E 3	https://openai.com/dall-e-3	網頁版（OpenAI）	透過文字敘述就能輕鬆生成圖像，操作門檻低且創作彈性大。
Adobe Photoshop	https://www.adobe.com/products/photoshop.html	桌面版（Windows/Mac）	以「生成式填充」等 AI 功能，支援自動修圖、擴充場景、合成內容，讓照片編修更便捷。

● AI 語音和音訊

工具程式	連結處	支援平台	補充說明
ElevenLabs	https://elevenlabs.io	網頁版	可將文字轉成多國語言,且極擬真的語音效果,也支援角色語音及客製化配音。
Suno	https://suno.ai	網頁版	能將廣告腳本或文案直接變成音檔,適用於行銷、廣播等多種用途。
Synthesia	https://app.synthesia.io	網頁版	AI 主播,輸入文字產出各國不同申調的語音,擬真度極高。

● AI 簡報應用

工具程式	連結處	支援平台	補充說明
Beautiful.ai	https://www.beautiful.ai	網頁版	可依照內容自動排版並提供多種模板,幫你快速完成專業簡報。
Pitch	https://pitch.com/	網頁版	為銷售簡報需求量身打造,提供 AI 投影片生成與即時協作功能。

● AI 網站建設者

工具程式	連結處	支援平台	補充說明
Wix	https://www.wix.com	網頁版	透過 AI 自動產生網站框架,再用拖放式編輯器輕鬆完成網站建置。
Framer	https://www.framer.com	網頁版	結合設計到前端的全流程,AI 協助快速生成互動網站原型。

● AI 聊天機器人

工具程式	連結處	支援平台	補充說明
Runway	https://runwayml.com	網頁版	提供影片後製、特效生成與剪輯的一站式解決方案,支援多種 AI 視覺處理技術。
Wondershare Filmora	https://filmora.wondershare.com	桌面版 (Windows/Mac)	結合 AI 與易用介面,可自動優化影片剪輯節奏、轉場特效等,迅速完成影片編輯。
Sora	https://openai.com/sora	網頁版	最新的 AI 影片生成工具,可根據文字或素材自動產生影片內容,應用範圍多元。

後記

寫在發行前
這本書是寫不完了……

在創作本書的過程中，深刻體會到生成式 AI 發展的驚人速度，幾乎無法追趕。一開始，我只是想撰寫一本關於影像生成工具的實用指南，但在短短幾個月內，平台與技術數度更新，讓書稿數次推翻重寫。

尤其是到了這個月（2024 年 12 月），OpenAI 的影片生成應用 SORA 正式上線，再次擴展了生成式 AI 的可能性。從文字到影像，從靜態到動態，AI 的創作能力已進入了全新的維度。在這樣一個不斷變化的技術浪潮中，我只能謙虛地承認，這本書註定是一個「寫不完的創作旅程」。

AI 人工智慧：從科學研究到生活日常

AI 人工智慧是一個被研究數十年的技術，但自 2022 年 12 月 ChatGPT 問世後，世人才真正對 AI 的可能性有了全新的認識。各種生成式應用如雨後春筍般湧現，百業皆在探討應用可能，從生產製造、資訊安全、產業垂直應用到營運管理，任何你能想到的領域，都有人在努力與 AI 建立連結。

這樣的變革，有人狂熱興奮，有人焦慮不安，當然也有人選擇無動於衷。常聽到「未來的世界將淘汰不使用 AI 的人」這樣的論述。作為站在應用最前端的一員，坦白說，我並不完全認同，甚至厭惡這樣的「數位落差霸凌」言詞。

AI 並不需要被學會，因為它將是日常

趨勢目前看來很難逆轉，AI 將從有形的工具變為無形的存在，深深嵌入生活中的每個角落。背後當然有 OpenAI、Google、Meta、Amazon、xAI 等企業的投入與科學家的研究，但對多數人而言，只要好好使用，就像吃飯喝水一樣自然即可。然而，我比較憂心的是數位落差的擴大與社會階層的鴻溝加深。我衷心期盼，科技不該成為統治或霸凌的工具，而應是縮短距離、促進平等的力量。

創作的心路歷程

這段旅程中，我們見證了藝術與科技交融的奇蹟。每一個新工具的誕生，都為創作者打開了一扇新窗。我們親身體驗過從 Midjourney、DALL·E、Gemini、Adobe Firefly、Stable Diffusion，到如今的 SORA，每個平台都帶來了全新的啟發，推動我們對創意表達的不斷探索。

創作過程中的每一個挑戰，從技術學習到版權倫理，都是我們重新思考「創作」意義的契機。AI工具越強大，我們越相信，真正具備藝術價值的，依然是 深厚的人文素養、敏銳的生活觀察與豐富的歷史文化知識。

對未來 AI 世界的期許

展望未來，我們期盼 AI 技術與人類創意的結合 能夠激發無窮的可能性。希望技術的進步，成為創作者的助力，而非壓力。

AI 應是想像力的延伸，而非創意的取代。

同時，我們也期待在技術飛速發展的背後，能夠同步建立完善的智慧財產權保護與倫理規範，確保每一位創作者的努力都被尊重與珍視。最後，感謝每一位讀者，無論你是技術愛好者還是藝術創作者，願我們攜手走進這個充滿無限可能的未來， Dancing with the Possibilities of Life。

Louis 2024/12/31

向大師致敬

511

國家圖書館出版品預行編目資料

人人都會 AI 繪圖：開啟斜槓人生金鑰匙，2000 件生
成作品＋完整提示詞 / 黃稟洲（Louis Huang）著 .
-- 臺北市：三采文化股份有限公司, 2025.02
　面；　公分 . --（iLead；17）
ISBN 978-626-358-464-8（平裝）

1.CST: 人工智慧 2.CST: 電腦繪圖 3.CST: 數位影像
處理

312.83　　　　　　　　　　113009851

suncolor
三采文化

iLead 17

人人都會 AI 繪圖

開啟斜槓人生金鑰匙，2000 件生成作品＋完整提示詞

作者｜黃稟洲（Louis Huang）

編輯一部 總編輯｜郭玫禎　主編｜鄭雅芳　美術主編｜藍秀婷

封面設計｜李蕙雲　內頁版型設計｜賴維明　內頁排版｜賴維明、莊馥如

發行人｜張輝明　總編輯長｜曾雅青　發行所｜三采文化股份有限公司
地址｜台北市內湖區瑞光路 513 巷 33 號 8 樓
傳訊｜TEL：（02）8797-1234　FAX：（02）8797-1688　網址｜www.suncolor.com.tw
郵政劃撥｜帳號：14319060　戶名：三采文化股份有限公司
本版發行｜2025 年 2 月 7 日　定價｜NT$780